Environmental Impacts of Globalization and Trade

Global Environmental Accord: Strategies for Sustainability and Institutional Innovation
Nazli Choucri, editor

Nazli Choucri, editor, *Global Accord: Environmental Challenges and International Responses*
Peter M. Haas, Robert O. Keohane, and Marc A. Levy, editors, *Institutions for the Earth: Sources of Effective International Environmental Protection*
Ronald B. Mitchell, *Intentional Oil Pollution at Sea: Environmental Policy and Treaty Compliance*
Robert O. Keohane and Marc A. Levy, editors, *Institutions for Environmental Aid: Pitfalls and Promise*
Oran R. Young, editor, *Global Governance: Drawing Insights from the Environmental Experience*
Jonathan A. Fox and L. David Brown, editors, *The Struggle for Accountability: The World Bank, NGOs, and Grassroots Movements*
David G. Victor, Kal Raustiala, and Eugene B. Skolnikoff, editors, *The Implementation and Effectiveness of International Environmental Commitments: Theory and Practice*
Mostafa K. Tolba, with Iwona Rummel-Bulska, *Global Environmental Diplomacy: Negotiating Environmental Agreements for the World, 1973–1992*
Karen T. Litfin, editor, *The Greening of Sovereignty in World Politics (1998)*
Edith Brown Weiss and Harold K. Jacobson, editors, *Engaging Countries: Strengthening Compliance with International Environmental Accords*
Oran R. Young, editor, *The Effectiveness of International Environmental Regimes: Causal Connections and Behavioral Mechanisms*
Ronie Garcia-Johnson, *Exporting Environmentalism: U.S. Multinational Chemical Corporations in Brazil and Mexico*
Lasse Ringius, *Radioactive Waste Disposal at Sea: Public Ideas, Transnational Policy Entrepreneurs, and Environmental Regimes*
Robert G. Darst, *Smokestack Diplomacy: Cooperation and Conflict in East-West Environmental Politics*
Urs Luterbacher and Detlef F. Sprinz, editors, *International Relations and Global Climate Change*
Edward L. Miles, Arild Underdal, Steinar Andresen, Jørgen Wettestad, Jon Birger Skjærseth, and Elaine M. Carlin, *Environmental Regime Effectiveness: Confronting Theory with Evidence*
Erika Weinthal, *State Making and Environmental Cooperation: Linking Domestic and International Politics in Central Asia*
Corey L. Lofdahl, *Environmental Impacts of Globalization and Trade: A Systems Study*

Environmental Impacts of Globalization and Trade
A Systems Study

Corey L. Lofdahl

The MIT Press
Cambridge, Massachusetts
London, England

©2002 Massachusetts Institute of Technology
All rights reserved. No part of this book may be reproduced in any form by any electronic or mechanical means (including photocopying, recording, or information storage and retrieval) without permission in writing from the publisher.

This book was set in Sabon by Interactive Composition Corporation
Printed and bound in the United States of America.

Library of Congress Cataloging-in-Publication Data

Lofdahl, Corey L.
  Environmental impacts of globalization and trade : a systems study / Corey L. Lofdahl.
    p. cm. — (Global environmental accords)
  Includes bibliographical references and index.
  ISBN 0-262-12245-6 (hc. : alk. paper)
  1. International trade—Environmental aspects. I. Title. II. Series.

HF1379 .L63 2002
333.7—dc21   2001044755

In Memory of
Alver Lofdahl and
Finlay Petrie

Contents

List of Figures   ix
List of Tables   xiii
Preface   xv
Acknowledgments   xix

1 Introduction: The Challenges   1
   1.1 Current Views   5
   1.2 Social and Natural Environments   10
   1.3 The Role of Computation   22
   1.4 Meeting the Challenges   27

2 Lateral Pressure Theory   29
   2.1 The Classics   31
   2.2 Clarifying Concepts and Contentions   35
   2.3 Lateral Pressure as Linkage Theory   40
   2.4 Lateral Pressure Extended   44
   2.5 Methods and Strategy   53
   2.6 Toward Robust Specification   58

3 Contextual Imperatives   65
   3.1 Geographic Analysis   68
   3.2 Time-Series Analysis   78
   3.3 Conclusion   95

4 Untangling Complex Linkages: Statistical Analysis   99
   4.1 Prevailing Contentions   99
   4.2 The Simplest View: Univariate Analysis   106
   4.3 The View Less Simple: Bivariate Analysis   111

4.4 The Complex View: Multivariate Analysis   116
4.5 Conclusion: Revising the Contentions   124

5 **Exploring Complexity: System Dynamics Analysis**   127
   5.1 Recalling Analytic Tensions   128
   5.2 Representing Analytic Tensions   131
   5.3 The Environmental Lateral Pressure Model   136
   5.4 Critical Inferences and Implications   149

6 **Conclusion: Implications for Theory, Methods, and Policy**   155
   6.1 Theoretical Extensions   158
   6.2 Methodological Advances   161
   6.3 Policy Consequences   164
   6.4 Next Steps   167

Appendix A   Complex Structures and Dynamics   171

Appendix B   Conditional Plot Analysis   177

Appendix C   Latitude and Longitude Analysis   189

Appendix D   TC × GNP Tests   191

Appendix E   Dynamic Model Equations   201

Appendix F   Nonlinear Relationship Analysis   211

Appendix G   Dynamic Equilibrium Analysis   219

Notes   221
Bibliography   227
Index   241

Figures

1.1   Nonlinear Causal Relationship   24
1.2   Feedback Loops   25
2.1   The Linkage Challenge   34
2.2   International Relations on War   41
2.3   International Relations on the Environment   45
2.4   Environmental Lateral Pressure   59
3.1   North and Choucri Ecological Profiles   71
3.2   Social Indicators of Development Ecological Profiles   72
3.3   1990 Carbon Dioxide (CO_2) per Capita   73
3.4   1980–1990 Deforestation   75
3.5   World Biomes by Carbon Content   76
3.6   Gross National Product (GNP)   80
3.7   Gross National Product (GNP) per Capita   81
3.8   Total Carbon Dioxide (CO_2) Output   83
3.9   Mean Carbon Dioxide (CO_2) per Capita   84
3.10 Total Imports   87
3.11 Trade Ratio   88
3.12 Total World Population   89
3.13 Total Population Growth   90
3.14 World Forest and Woodland Area   93
3.15 World Agricultural Land   95
4.1 G_i^* GNP per Capita   109
4.2 G_i^* CO_2 per Capita   110

x Figures

4.3 G_i^* Population Growth Rate   111
4.4 G_i^* Deforestation Rate   112
4.5   Example Trading System   118
4.6   Trade Connected GNP   120
4.7 G_i^* Trade Connected GNP   121
4.8   Competing Model Perspectives   126
5.1   Overshoot and Collapse System Structure   132
5.2   Overshoot and Collapse Feedbacks   133
5.3   Overshoot and Collapse Dynamics   135
5.4   Environmental Lateral Pressure (ELP) System Structure   138
5.5   Resource Trade System Structure   139
5.6   Technical Trade System Structure   140
5.7   Population and GNP Feedbacks   141
5.8   GNP and Resource Feedbacks   143
5.9   Trade Feedbacks   144
5.10 North Dynamics without Trade   145
5.11 North Dynamics with Trade   147
5.12 South Dynamics with Trade   148
5.13 Environmental Lateral Pressure (reprise)   150
A.1   Fractal Structure   172
A.2   Algebraic Curves   173
A.3   Chaotic Curves: Lorenz Attractor   174
A.4   Chaotic Curves Compared: Duffing Oscillator   175
B.1   Profile Pattern of Conditional Plots   178
B.2   GNP by Profile   178
B.3   GNP per Capita by Profile   179
B.4 CO_2 by Profile   180
B.5 CO_2 per Capita by Profile   181
B.6   Imports and Exports by Profile   182
B.7   Trade Ratio by Profile   183
B.8   Population by Profile   184
B.9   Population Growth by Profile   185

B.10 Forest and Woodland Change by Profile   186
B.11 Agricultural Land by Profile   187
D.1   Forest Change Model Diagnostic Plots   192–193
F.1   Overshoot and Collapse Converter Graphs   212
F.2   Population Converter Graphs   212
F.3   GNP Converter Graphs   213
F.4   Resource Converter Graphs   214
F.5   Resource Trade Effect Converter Graph   215
F.6   Technical Trade Converter Graphs   216
G.1   South Dynamics without Trade   220

Tables

3.1 Ecological Profile Definitions   69
3.2 1991 Ecological Profile Groupings   70
4.1 Absolute Latitude as a Function of the Four Major Variables   107
4.2 CO_2 per Capita $= f$(GNP per Capita)   113
4.3 CO_2 per Capita $= f$(Population Growth)   114
4.4 Forest Change $= f$(Population Growth)   115
4.5 Forest Change $= f$(GNP per Capita)   116
4.6 Forest Change $= f$(TC $\times$ GNP)   122
4.7 Forest Change $= f$(TC $\times$ GNP + GNP per Cap. + Pop. Growth)   123
5.1 Comparative Advantage Example   129
D.1 Panel Corrected Standard Errors (PCSE) Test   197
D.2 Cross-Sectional Tests   198
D.3 Lagged Spatial Weight Tests   200

Preface

Within the field of international political economy, questions of environmental degradation are seldom addressed. Yet the degradation of nature can be one of the most pervasive and longest lasting consequences of development. Too often, such consequences are simply catalogued without sufficient consideration given to their social—that is, political and economic—causes. This shortcoming is addressed by examining the complex relationships among globalization, trade, and the environment.

To ensure that the study is focused and organized, a roadmap, plan, or vision is required. This is partially accomplished by addressing a substantial, open policy question: Does international trade help or hurt the environment? The Seattle riots at the World Trade Organization (WTO) meeting in November and December of 1999 emphasize the increased salience of this issue. Economists maintain that trade helps the natural environment as rich countries can better afford to protect their unspoiled areas. Environmentalists counter that the pursuit of national wealth drives global environmental degradation and that free trade accelerates this process. The goal of this study centers on resolving the conflict between economists and environmentalists concerning trade.

A topical and interesting question can only provide so much direction. Theory is used generally to establish analytic boundaries and direct research. Lateral pressure theory is used specifically to account for the politics and economics within and among nations. Whereas lateral pressure has traditionally been used to explain war (Choucri and North 1975; Choucri, North, and Yamakage 1992), this study extends the theory's initial application to the environment (Choucri 1993a) by specifying additional connections between the social and natural environments. Doing so requires analytic, methodological, and empirical support for several reasons. First, it is understood that theories divorced from empirical measurement are doomed to sterility and irrelevance. Second, it is equally understood that observations without theories to organize them tend to confuse more often than they clarify. Third, appropriately chosen methodologies ensure that theory and data are connected in useful, defensible, and comprehensible ways.

Faced with the challenge of combining theory, methodology, and policy in creating a coherent analysis, where does one turn for guidance? This study is organized according to the system dynamics *reference mode* (Randers 1980), which has proven useful in ordering inherently unstructured problems. The reference mode concept, as used herein, consists of five parts: (1) identify the timeframe and major themes of the study, (2) describe and develop the theory that motivates and organizes the study, (3) identify primary variables and graph them over time, (4) identify sign and magnitude of the causal connections among the variables, and (5) develop a simulation model based on the previously specified timeframe, themes, theory, variables, dynamics, and connections.

The first five chapters correspond to the five parts of the reference mode. Chapter 1 introduces the key terms: globalization, trade, environmental impacts, and system—using the Seattle WTO riots as a point of departure. Implicit throughout this study is the divide between the social and the natural environments, which manifests itself in three ways: (1) an ideologically divergent globalization literature, (2) the separate concerns and histories of the social and natural sciences, and (3) the impact of increasingly potent analytic methods and tools. Generalizing broadly, the social sciences provide the most persistent and interesting questions, while the natural sciences provide the most advanced analysis tools. Social insights and natural methods combine to address global environmental degradation because, although the problem is natural, the solution must be social. This tension between the social and natural is addressed in terms of politics and markets, geopolitics and power, and computational advances. Chapter 2 strengthens the theoretical foundation by extending lateral pressure so that it can be used to explore environmental questions. Whereas standard social science theories are spare and static, lateral pressure is comparatively complex and dynamic.

The first two chapters provide the theoretical foundation for the subsequent empirical analysis. Throughout the following three chapters, historical data is analyzed in fundamentally new ways. To understand how this analysis differs from previous efforts, consider that economist's attempts to measure the environment in terms of wealth or utility have not provided much analytic benefit (Economist 1998), in large part because figuring out the price people are willing to pay for environmental perquisites does not yield much analytic traction. What is required is a more multidimensional fashion of representing complex systems, one that allows analysts to value separate things separately and yet still connect and compare them in defensible and illuminating ways. Striving to keep the multiple indicators of a complex system in balance rather than maximizing a few values results in more sustainable and mature policy prescriptions.

Complexity brings with it a strong orientation towards system, which implies the social sciences need to borrow analytical techniques from the natural. It is not enough to recognize that a problem is "complex" and then stop. Thus a significant problem is addressed in a manner that illustrates just what complexity means. As interpreted here, complexity implies a more systemic, all-encompassing, and cross-national analysis as opposed to case studies. Chapter 3 introduces the variables of interest and displays their dynamics over time to give a general feel for the problem at hand. Chapter 4 sharpens the study's focus by establishing empirically supported causal connections among the variables of interest, especially those linking trade and the environment. The debate between economists and environmentalists is addressed most explicitly here. Chapter 5 connects the theoretical portion of the analysis with the empirical by making explicit the complex—that is, multiple and nonlinear—causal connections that exist between world trade and global environmental degradation.

Chapter 6 provides an opportunity to summarize and extend the lessons learned throughout the study. The major finding is that trade and environment are connected: Gross National Product (GNP) increases in the rich, developed countries are linked to deforestation in the poorer, developing countries. Beyond this though, a range of other conclusions are presented. With respect to theory, lateral pressure is extended and applied to explain environmental degradation by comparing and contrasting local and global. However, the challenge of capturing, developing, and empirically defending such explanations is just as methodological as it is theoretical. Data visualization tools, statistical models, and computer-based simulations combine to handle the complexity of globalization by ordering observations and rendering them understandable. As regards policy, the simulation model presented herein was kept simple by design to highlight more theoretical insights and improve understandability; thus it is too simple to be used for rigorous policy development. However, the insights derived from this study are sufficient to critique some of today's current trade policy prescriptions. Choucri (1981) provides an example of a large and complex modeling effort used to study the world economy—specifically, international petroleum markets. A similar effort would be necessary to develop policy prescriptions regarding trade and the environment.

Finally, the motivation for this effort is addressed. At a personal level, I wanted to apply my early education in electrical engineering and computer science to an important question. Some have asked why, with a technical background, I chose to study international relations. The answer is, while the problems of global environmental degradation are essentially physical, their solution has to be political. To study this topic then, one must understand both the social and natural environments. This intuition suffuses this work, and while an important environmental policy question is indeed addressed, another, larger project is also touched upon:

> The greatest enterprise of the human mind always has been and always will be the attempted linkage of the [natural] sciences and the [social] humanities. The ongoing fragmentation of knowledge and resulting chaos in philosophy are not reflections of the real world but artifacts of scholarship. The propositions of the original Enlightenment are increasingly favored by objective evidence, especially from the natural sciences. (Wilson 1998, 8)

"Enlightenment" here implies the assumption of a lawful material world, the intrinsic unity of knowledge, and the potential for indefinite human progress. Environmental policy is the door that grants entry to the debate, and the analysis of globalization, trade, and environment opens the door.

Acknowledgments

I would like to thank CIESIN for making the Social Indicators of Development dataset available online. Jani Little proved invaluable in helping me to access the IMF Direction of Trade data. Oak Ridge National Labs was most helpful in providing me with their CO_2 and biome data. Thanks also to Eric Schol for creating the fractal graphics.

Studying and writing in two cities leads to the happy consequence of having two sets of colleagues. In Boulder, Colorado, I thank Steve Chan, Michael Glantz, Horst Mewes, and John O'Loughlin. Much appreciation goes to the University of Colorado Department of Political Science, the Center for International Relations, and to my colleagues at the Institute of Behavioral Science's Program on Political and Economic Change for their kind assistance and support including Mohan Penubarti, Jordin Cohen, Ed Greenberg, Walt Stone, James Scarritt, Tom Mayer, David Brown, and David Reilly. Michael Shin provided much assistance with ARC/INFO and other mysteries, Kristian Gleditsch helped greatly with the panel corrected standard errors analysis, and the National Science Foundation (SBR-9511577) funded my initial forays into diffusion modeling.

In Cambridge, Massachusetts, I thank Hayward Alker, Lincoln Bloomfield, and Harvey Mansfield for their advice and counsel. John Sterman, Jay Forrester, Nan Lux, and those at the MIT System Dynamics Education Project were instrumental in my learning system dynamics. Clay Morgan, with the help of multiple anonymous reviewers, gently guided this research towards publication. Special thanks also goes to Jette Knudsen, Jan Sundgren, Chris Heye, Dan Lindley, Avik Roy, Steve Wilmsen, Alan Evans, Mike H., all at SAIC, and Elizabeth McLaughlin who is missed very much.

Finally, I thank Michael Ward and Nazli Choucri for giving me the opportunity, guidance, and support necessary to complete this study. Robert North merits special mention for his vision and inspiration. To my family—Lisa and Lisa, Mom and Dad—my accomplishments, academic and otherwise, would not have been possible without your many sacrifices unselfishly made on my behalf.

Environmental Impacts of Globalization and Trade

1
Introduction: The Challenges

In autumn 1999, the World Trade Organization (WTO) met in Seattle to negotiate a new round of talks that would further liberalize international trade. What actually happened deviated significantly from the plan. The meeting will be remembered not for a new round of trade talks, which did not happen, but for the antitrade protests that did. Trade meetings have traditionally generated only polite interest in their host cities, but in Seattle, WTO delegates were greeted by large, intense, well-organized, and sometimes violent protests against globalization. There were protesters, thousands of them dressed in colorful and imaginative costumes, carrying banners and protesting peacefully through the streets. There were trade delegates who were baffled by the attention and prevented from getting to their meetings by the protests and other forms of resistance. Some protesters chained themselves together to form human barriers, others smashed shop windows and looted businesses, while still others battled police who made liberal use of tear gas and truncheons. The images of passion, demonstration, and confrontation in Seattle were transmitted around the world where they made an impact on popular perception and public policy.

The protests were not directed at the trade delegates, government officials, economists, and corporate executives as people but as representatives of globalization. The protesters were not angry with globalization per se but with the injuries they perceive were inflicted by it on the world's poor and the global environment. The WTO delegates considered the Seattle protesters fundamentally misinformed and misguided. Proponents of trade believe, in contrast, that globalization and trade help the poor and the environment. Such opinions are not held lightly, supported as they are by decades of economics scholarship showing that trade barriers lead to depression, while trade liberalization leads to economic growth, arguably the best mechanism for lifting populations out of poverty. The WTO, working as it does to lower tariffs and increase international trade, in its opinion helps the very people about whom the protesters profess to care.

The WTO's views are not shared by the antiglobalization movement. The movement's agenda is comparatively tough to pin down however, composed as it is of so many different groups, causes, and opinions. Broadly speaking through, there were three major groups in Seattle. First were the young and disaffected who were there to fight authority, and who might describe themselves variously as anarchists, activists, Marxists, students, or simply predisposed to violence. These people were present and garnered a disproportionate share of attention. Second were the unions who were there to protest lost jobs, lower wages, and reduced labor standards. In the unions' view, trade reduces jobs when products that were made domestically are instead produced abroad and then imported. Such goods are produced abroad more cheaply when foreign workers work for less than domestic workers. Even if goods remain domestically manufactured, the very threat of less expensive foreign production works to drive down domestic wages and labor standards.

The third group of protesters, and arguably the most central and voluble, were the environmentalists. At first, the relationship between trade and the environment is not obvious. Linkages between trade and the environment are roughly the same as those between trade and labor issues. Just as foreign goods can be made for less when labor standards are ignored, so too can they be made for less when environmental standards are ignored. No longer an academic question, the relationship between trade and the environment has become an increasingly contentious issue for economists and environmentalists. Several times over the past decade, trade disputes have been brought to the WTO and its predecessor, the General Agreement on Tariffs and Trade (GATT) (i.e., when the United States unilaterally stopped imports for environmental reasons). Specifically, the United States disagreed with the environmental damage caused in making certain products.[1] Such actions have been disallowed because it is a fundamental principle of international trade that one country cannot stop imports from another because it disagrees with the way those products were produced, and this includes environmental impacts. Should a developing nation choose to incur more environmental damage in the production of its products, then that is its prerogative, its *comparative advantage* (cf. section 5.1). Trade economists argue that a country making such a choice should not be penalized for doing so. Environmentalists, who value the natural environment over economic efficiency and international trade, disagree with such reasoning.

The meaning and long-term consequences of the Seattle protests remain unclear. The 1999 WTO meeting did not produce a new round of trade liberalization talks as intended. Whether that was due to the protests or to the long-festering schisms between developed and developing nations has not been determined. Such disagreements stem from the developed nations desiring higher labor standards in the developing nations and the developing nations demanding access to the markets of the developed, especially for products that benefit from low wage-rates like textiles and agricultural goods. Some predicted that trade protests would prove ephemeral, an aberration unique to Seattle, but this has not been the case. Similar protests have occurred at subsequent meetings of the International Monetary Fund (IMF), the World Bank, and the World Economic Forum (WEF).[2] Clearly the protests represent something significant, but what?

Rather than contrast and compare the various protests' methods and members, let us instead consider the question of what drives the policy impasse between trade's proponents, the economists, and their opponents, the environmentalists. Doing so attempts to tease out the inchoate and unspoken differences between economists and environmentalists by articulating the fundamental beliefs held by each. As has been noted, economists maintain globalization and trade help developed and developing nations grow economically, which in turn brings much needed capital, investment, and opportunity to the world's poorest populations. Environmentalists, in contrast, argue that the increased economic activity among nations caused by globalization and trade leads to a lowering of regional labor standards and an unhealthy linking of developing countries to their more developed trading partners. From the human point of view, there is a fundamental conflict over the consequences of globalization for the world's poor: Does globalization help or hurt?

The Seattle protests were not driven solely by concern for the anthropocentric consequences of globalization, although that was certainly part of it—just as important were the consequences for the natural environment. While environmentalists were certainly irked by the WTO's history of environmentally contrary rulings, it is unclear if that issue alone would have motivated such a large and diverse collection of protesters. It is at least arguable that forces more fundamental than bureaucratic unresponsiveness motivated the protests, namely that the environmentalists believed that a strong and fundamental relationship exists between globalization and trade and also between trade and global environmental degradation. Put simply, the protesters maintain that trade hurts the environment. This belief is in direct contradiction to economists who hold that trade helps the environment. Just before Seattle, the cover of *The Economist* (1999a) offered to explain "Why Greens Should Love Trade." The logic rests on the fact that rich, developed countries have better environmental conditions than their poor, developing counterparts. Economists go on to argue that this relationship exists because rich countries have the intellectual and financial resources necessary to protect their environments. Because trade contributes to national wealth, trade consequently provides the resources required for poor countries to protect their environments as well.

Reconciling these competing views—that trade either helps or hurts the natural environment—is a challenge. But addressing, analyzing, and answering this challenge brings with it a host of other challenges. As has been noted, the impasse between economists and environmentalists is being addressed in Seattle and a host of other cities in a manner that will leave the participants politicized, polarized, physically injured, and unenlightened. Having one side of the debate working to increase trade while the other battles police to stop it is no way to decide a complex policy question. This study instead addresses the environmental impacts of globalization and trade using the tool of quantitative analysis. Instead of arguing for one side or the other, this study seeks to uncover and articulate the logic supporting each side as a way to reconcile the parties and point the way toward more productive policies. This chapter pursues this goal in three sections. First, the recent literature is reviewed to reveal the current debate about globalization, trade, and the environment. Second, the intellectual histories of international political economy and geography are presented to reveal the historical context in which economists and environmentalists operate. Third, recent advances in quantitative analysis are reviewed in order to understand their impact on the debate. In conclusion, the question "Does trade help or hurt the environment?" is revisited and revaluated in light of the previous discussion.

1.1 Current Views

Globalization serves as the starting point for any current analysis of the international political economy, whether one is for or against it (Hirst and Thompson 1999). The term *globalization* itself is defined variously, its scope growing and contracting according to the needs of the moment, but fundamentally the term implies increased linkages across national boundaries, expansion of the international market economy, and a complex and integrated world society (Lechner and Boli 1999). Within this study, globalization is operationalized in terms of trade, while other aspects of international interaction are captured by the globalization concept including Foreign Direct Investment (FDI), Multi-National Corporations (MNCs), advanced communication technologies, and migration (Economist 1997b).

The basic question here is not to define globalization but to determine its consequences. Is globalization on the whole a positive or negative force in the world, and who thinks so? Is globalization an engine of prosperity and wealth for the majority of the world's population or a source of inequality, instability, unemployment, and environmental degradation (Baker, Pollin, and Epstein 1999). Such questions are examined first by looking at the arguments supporting globalization that derive primarily from the business management and economics literatures. Next the arguments against globalization are examined. This is a less well defined, less coherent, and more descriptive literature than its economics-based counterpart. In conclusion, some lessons will be drawn from these two divergent literatures, the major one stating that neither completely free markets nor centralized command and control economies are the answer to the ills of globalization. Some hybrid between the two will ultimately prove necessary, but just how that might happen remains an ongoing opportunity.

Globalization does not just happen, people make it happen. Thus there is a significant proglobalization literature that not only describes globalization's macrolevel paths and processes but also illustrates the microlevel strategies required to survive and thrive within its throes. Business books in particular propound the thesis of adaptation, competition, and survival. Isaak (2000) studies how to manage change in the global economy by stressing its human, strategic, and political dimensions. Marquardt (1998) offers a recipe for global success by focusing on corporate culture, human resources, strategies, operations, structure, and organizational learning of best practices. Examples are drawn from many of the world's must successful and valuable companies including General Electric, Whirlpool, Colgate-Palmolive, Shell, Coca-Cola, Xerox, Federal Express, and Hewlett-Packard, but the basic message is that companies must either globalize or die. Ohmae (1999) seeks to change the way managers invent, market, and compete in the new, global marketplace by challenging their assumptions with provocative observations. Ohmae suggests that companies must serve customers, not governments; wealth comes from markets, not natural resources; and national borders are irrelevant. Whether or not such statements are true is less important than convincing managers to think in global terms for the competitive good of their companies.

The defense of globalization does not derive solely from those who compete within markets; more economically informed works also defend it. Burtless et al. (1998) argue that Americans have historically benefited from globalization as trade delivered cheaper and better foreign goods, fostered improved international relations through the export of democracy, and even helped the United States win the Cold War. Since the end of the Cold War though, globalization's detractors have become more vocal. Burtless et al. argue, however, that globalization should be pursued still further because the pain felt by working Americans, while real enough, has not been caused by globalization. Policies need to be created to help such people, and backing away from the economic benefits provided by trade is not the answer.

Friedman (1999) provides another defense of globalization using the metaphor of a Japanese car, the Lexus, to symbolize the modern world of globalization and consumerism. Friedman contrasts the Lexus image with that of the Middle Eastern olive tree, which represents the old world of religion and thick, family-based culture. In this way, Friedman portrays globalization as a progressive and explicitly transformative project, one that has come to be the major organizing principle of the international system over the last fifty years. According to Friedman, globalization becomes less of a competitive or economic question than a philosophical one that centers on whether individuals choose their lifestyles for themselves or whether society determines lifestyles for the individual. The answer, according to Friedman and globalization's proponents, is a resounding yes for individual choice. People choose the Lexus. People want choice. Insofar as this process has taken hold around the world, it is often perceived as creeping Americanism. Friedman disagrees. Globalization is not so much the result of American power as the natural and progressive development of international relations, cultural institutions, and the rise of individualism.

A new strain of progressive international economic scholarship has emerged that highlights the unintended consequences and costs of globalization from the point of view of poor, developing countries. For example, Hoogvelt and Popova (1997) investigate emerging, global forms of production, exchange, and governance from the perspective of Sub-Saharan Africa, the Middle East, East Asia, and Latin America. Korten (1998) concurs that the poor in these regions experience significant unmet needs including hunger, housing, unemployment, poverty, human rights, and the degradation of their natural environment. Shiva (1997, 1999) takes a more provocative view of globalization's disparities maintaining that modern, biotechnical discoveries drive the continued exploitation of women, plants, and animals. She goes on to argue that the WTO effectively delivers developing countries to Western companies in a modern form of colonialism.

Companies with global reach or Multi-National Corporations (MNCs) are powerful international actors that further globalization and cause many of its significant negative consequences. Several characteristics of MNCs support this conclusion. First, ranging as they do across national borders, MNCs are unbounded by national loyalties and are free to pursue short-term economic gain, which often comes at the expense of the natural environment (Karliner 1997). So strong are MNCs that in some nations they, and not the local governments, control the economy (Barnet and Cavanagh 1995). Cavanagh (1992) notes that, while MNCs may promise all manner of benefits, they can only reliably be counted on to deliver power to their leaders. The North American Free Trade Act (NAFTA) which, according to Cavanagh, features environmental, agricultural, human rights, and immigration abuses, is offered as an example. Danaher (1997) points out that economic opportunity is actually shrinking for Americans, especially for minority women, due to the declining wages offered by MNCs. Korten (1996), after noting similar problems, argues for development that serves people rather than the narrow interests of MNCs.

Others look not to poor nations or MNCs but to globalization itself, describing it as a large-scale, international reordering on the scale of industrialization. Mander and Goldsmith (1996) point out that the future of a globalized world may not be as bright as its proponents would have us believe. For instance, Jameson and Miyoshi (1998) investigate globalization's far-reaching impacts on technology, communications, consumer culture, intellectual discourse, the arts, and mass entertainment, once again making the point that globalization is as much a cultural force as it is economic. More specifically, Brecher and Costello (1998) argue that international competition for jobs by labor leads not so much to a "race to the bottom" as standards erode but a "free fall to the bottom" that benefits primarily the wealthy. Consequently Rodrik (1997) predicts a backlash against globalization in the form of labor unrest in the United States, Europe, and Asia; he further predicts that globalization will become the next great foreign policy debate. Implicit in these predictions is the recognition that globalization is a socially destabilizing force, both at the personal and national levels (Greider 1997). Moreover, as oversupply, cultural dislocation, and threat of depression increase, it must be recognized that such forces are beyond the ability of any regulatory institution to control. This raises the threat of trade barriers and protectionism implemented by the economically disadvantaged (Rodrik 1997) or the reassertion of government control over markets (Amin 1997).

Gray (1999) goes further in arguing that international free-market economics as currently practiced is antithetical to social stability and environmental protection. In making this point, Gray draws a distinction between globalization, which he defines as inevitably increasing international connections, and global capitalism, an ever-expanding version of Anglo-American market capitalism. According to Gray, global capitalism is essentially an American political project. So powerful and politically toxic is global capitalism that it threatens liberal civilization by causing uneven economic development, structural unemployment, and the fall of real wages. If left unchecked, global capitalism will lead to an economic endgame of trade wars, competitive currency devaluations, economic collapse, and political upheavals. Gray contends that international economic competition also contributes to environmental degradation. Developed countries protect their domestic environments by moving production to developing countries where environmental standards are looser, which effectively exports pollution.

Criticisms of the international economic order are not new. The WTO riots in Seattle is an American articulation of the deep-seated frustrations with globalization that much of the developing world has long felt. What emerges from Seattle and the current globalization literature are two quite divergent schools of thought—one strongly supportive of globalization and the other just as strongly opposed. Since globalization is likely to persist in some form for the foreseeable future, some accommodation between the two camps is required and desirable. Baker, Pollin, and Epstein (1999) recognize that globalization is a force, for better or worse, depending on the type and degree of regulation that governments are able to impart on the trajectory of their development path. Dunning (1999) concurs, noting that the increased power of MNCs and the daunting complexity of globalization present significant policy challenges to governments. When reevaluating governments' roles and responsibilities, Rodrik (1999) stresses sticking to fundamentals including democracy, openness, and integrity. Kuttner (1999) emphasizes the ongoing need for regulated markets, citing America's experience with employment, medicine, banking, securities, telecom, and electric power among other industries. Some regulation of the free-market forces unleashed by globalization is necessary, both from a labor and environmental perspective; just what form such regulation might take and what role the WTO might play in its formation remains an open question. While the current literature has been reviewed, it provides only a snapshot of the present debate. Understanding globalization's larger themes and underlying issues requires a review of its institutional and intellectual history. The following section addresses these latter concerns.

1.2 Social and Natural Environments

In determining the best course of action for government in the emerging global context, it helps to understand the thoughts, concepts, and policies that have led to today's state of international affairs. This section provides two ways of understanding international relations and their sometimes complementary and contradictory intellectual histories. On the proglobalization side, a short history of the *social environment* is presented in terms of politics and markets; on the antiglobalization side, a history of the *natural environment* in terms of geopolitics and power is presented. One of the key differences developed in this section turns on the topic of systemic and analytic complexity. Politics and markets are grounded in the field of economics, arguably the premiere social science, and yet its greatest policy successes rest on fairly simple concepts. In contrast, geopolitics and power are grounded in the field of *geography*, a discipline that recognizes the inherent complexity and interconnectedness of social and natural environments. Despite this important observation, geography remains less influential than economics, although recent computational advances are breathing new life into the discipline by increasing the transparency of complex systems.

1.2.1 Politics and Markets

The intellectual foundations of globalization and trade were developed by two forefathers of modern economics, Adam Smith and David Ricardo. Their ideas collectively form the foundation for the twentieth century's global economy. To cover the full scope of their influence is beyond the bounds of this study, but three key concepts are addressed: (1) development, (2) growth, and (3) trade. It is argued that these fundamental concepts, ideas that have for so long guided the development of politics and markets, are now being called into question as their true costs, social and natural, become increasingly apparent. The certainty and clarity that initially gave their ideas such force and power are being challenged by more critical, competing views.

Underlying the whole of modern political economy is the notion of *development*. Its original formulation derives from Adam Smith's recombining of the economic policies of his day. Eighteenth-century English economics featured the interests of pro-industry mercantilists and pro-agriculture Physiocrats. The mercantilists' valuation of industrial wealth was accepted by Smith (1937 [1776]) while their policy of nationalist economic interference was rejected, and the Physiocrats' policy of free trade was accepted while their valuation of agricultural wealth was rejected. So development, as envisioned by Smith, is both industrial and trade-based. Moreover, development implies increased productivity from specialization and division of labor, and this in turn entails social reorganization as the economy changes from an agricultural to an industrial base. While it is recognized that such reorganization may be stressful and painful for those involved, it is a necessary part of development and of creating civilization out of nature, which is inherently good (Smith 1869 [1759], 162–163; Cropsey 1987a, 645–646).

Growth implies the increase or expansion of some thing or process. When growth comes at the expense of something else, it implies a transition from that which came before to that which is to come after (Prigogine 1980). The transition from an agricultural to an industrial society depends on the growth of manufacturing and technology, a policy pursued based on both its simple appeal and theoretical promise. Much of development's appeal, however, is based not on pure economics but on the political exigencies of eighteenth-century Europe. The division of labor entails the creation of markets in which the ambitious and talented are able to ply their trade and sell their wares in a competitive arena that tests their mettle in a manner other than warfare. This concept was radical and even revolutionary when contrasted with the staid, semifeudal agricultural societies of the day in which loyalty to the manor's lord still mattered above all else, an era that provided no outlet for the ambitions of those dissatisfied with their station. Smith's economics provided a way to channel this energy and ambition toward constructive ends through market economics (Cropsey 1987a, 642; Samuelson and Nordhaus 1992, 3). The price and wage information provided by markets helps to transform personal dissatisfaction into goods and services for which people are willing to pay. Growing markets allow for ever greater specialization and ever growing opportunity to provide services, start new ventures, and accumulate personal wealth. Thus, Smith's famous "invisible hand" transforms heretofore problematic personal ambitions into socially beneficial behaviors—contributions to the general wealth or gross national product—although the social benefits are essentially unintended (Cropsey 1987a, 645–646, 650; Samuelson and Nordhaus 1992, 376).

Ricardo, however, considered the long-term consequences of industrial growth and recognized that capitalism's "golden age" as an engine for social progress would last only so long as labor was valued more highly than capital or natural materials (Samuelson and Nordhaus 1992, 377, 548–549). Ricardo reasoned that as population continued to grow and land area remained constant, wage rates would eventually be driven down to subsistence levels as all available land became occupied (Ricardo 1911 [1817]; Pearce 1992, 377–378). *Trade* provided the solution to this dilemma in two ways. First, importing cheap foreign food allowed England to specialize in high-value manufactured goods instead of land-intensive agricultural products. Second, Ricardo's doctrine of *comparative advantage* shows that specialization through international trade proves an economic boon for any country that participates (see section 5.1). Reallocating workers to make products in which one country has a comparative advantage over another allows for greater consumption in both countries (Samuelson and Nordhaus 1992, 663–671). Ricardo argues for free trade just as did the Physiocrats and Adam Smith, but comparative advantage provides a more concrete, analytic rationale for doing so. Moreover, the manner in which Ricardo framed comparative advantage showed economists specifically and intellectuals generally how to offer important policy recommendations based on simple models with large assumptions, a practice that remains widely popular today (Pearce 1992, 278).

Smith and Ricardo's ideas regarding development, growth, and trade formed the basis not only for neoclassical economics but also for the structure of the modern international system. The first country to implement their concepts was Great Britain, which dominated the international system throughout the nineteenth century (Roberts and Roberts 1980, 599–665). Britain's role as leader of the world economy came under challenge by Germany and the United States in the early 1900s.

World War I brought the challenge to the fore, and by the war's end Britain's time as the world's leading power was over. The United States was Britain's natural successor, but it was reluctant to assume the role. U.S. President Woodrow Wilson blamed World War I on the power-based politics practiced by Great Britain and other European powers, and he did not want the United States to contribute to its continuation, even if it got to lead. He threw his energies instead into the League of Nations that was created to provide international leadership through collective security. It was originally envisioned that the League would identify aggressor states and that the collective strength of the participating states would be brought to bear on these aggressors, thus preventing war (Kegley and Wittkopf 1995, 522–523). The presence of the League and the absence of a single state leader proved unsuccessful. Economically, the absence of effective economic coordination proved disastrous as the world trading system devolved into nationalist protectionism and worldwide depression (Kindleberger 1973, 172). Economic problems exacerbated national difficulties that in turn created restless populations ready to receive political entrepreneurs and their tales of imperialist salvation. Politically, the League of Nations failed to retard the aggression of a rogue state, Nazi Germany, leading to the abandonment of collective security as an international policy tool (Carr 1964 [1939]).

World War II resulted in part from an excess of political centralization within the League coupled with an absence of economic leadership. This set the stage for the Bretton Woods Agreement of 1944 forged by Lord Keynes of Great Britain and Dr. Harold White of the United States, the meeting at which the International Monetary Fund (IMF) and the International Bank for Reconstruction and Development (IBRD), or World Bank, were born. These institutions, along with the General Agreement on Tariffs and Trade (GATT) created in 1947, formed the basis for the postwar Liberal International Economic Order (LIEO), an era marked by open and decentralized market relations maintained through American leadership. This transformation was driven by the belief that free trade and open economic systems are the best way to promote economic growth and avoid war (Carr 1964 [1939], 235–239). Between 1950 and 1985, the world's population doubled from 2.5 to 5 billion while the world's output and trade increased more than five times their 1950 values (Chisholm 1990, 94–95). The LIEO dominated the postwar international system by delivering economic growth, albeit with challengers and changes along the way.

The Soviet Union challenged America's leadership from the end of World War II until it collapsed in 1989. Although the Soviet Union and United States never fought each other directly, a constant state of threat and tension, or Cold War, persisted between the two presenting both an economic and military threat to the LIEO. The economic challenge was one of a centrally controlled economy set against a market economy, with the communist or Marxist nations led by the Soviet Union practicing the former and the industrialized nations led by the United States practicing the latter. The Cold War period was also one of decolonization and nationhood for much of the world, the number of states increasing from 76 in 1946 to 157 in 1994 (Jaggers and Gurr 1995). Most of these states, however, did not enjoy the same economic success as their more established and developed counterparts, and this led to yet another international division, that between the industrialized states to the north and the developing or Third World states to the south (the first world being the leading LIEO nations and the second the communist). The economic disparity between the developed North and developing South led to calls for a New International Economic Order (NIEO) that would redistribute wealth to the poorer countries in compensation for colonialism among other reasons. The postwar incarnation of world government, the United Nations, became the forum for such demands as the poorer countries far outnumbered the wealthy.

The United States thus finds itself in a curious position at the Cold War's end. It is the world's peerless military power. Economically, it remains very strong if not exactly dominant the way it once was. Ideologically, though, the victory of LIEO is not so clear (Fukuyama 1989, 1992; Brzezinski 1993). Just because liberal capitalism emerged victorious at the end of the twentieth century does not mean it is the best possible form of political and economic organization as its long-term stability and sustainability have increasingly come under question. Socially, will liberal capitalism continue to satisfy the wants and needs of modern individuals, or will its victory prove ephemeral as the world descends again into poverty and war? Environmentally, will the natural environment be able to provide enough raw materials to sustain the LIEO, or will nature constrain and rescind these hard won economic achievements?

Such questions constitute a direct challenge to economics as a discipline. Economics is among the most credible, influential, and methodologically grounded of the social sciences. Economists have delivered on their promises. Other disciplines have yet to establish a comparable record of success, and so those who question economics—such as the anti-WTO protesters in Seattle—are regarded skeptically by those in the international economic policy community. Such skepticism carries over to the training and thinking of those entering the field. Economists, when learning the basic theories of their discipline, have ample opportunity to reflect on the development, history, and utility of each theory. It is true that some ideas work out, some don't, and some need to be modified before they can make a contribution. The field's success gives economists ample opportunity to develop confidence in the foundations of their discipline as well as ample examples to deflect specious attacks on their methods. Antitrade protesters and environmentalists do not have a comparable body of experience or proven analytical techniques from which to draw. Until they solve a major, open research question that has gone unanswered by traditional economics, their credibility will continue to be questioned and their concerns will continue to be overlooked.

Looking directly at the economic theories reveals issues beyond topical focus and key concerns. The theories of neoclassical economics carry over from the discipline's history, which is to say they are simple. Reducing complex problems to a small number of variables makes them relatively easy to communicate and analyze. While such models are not perfect representations of reality, they have proven good enough to support the creation of the global economy, a considerable achievement in its own right. Antitrade protesters and environmentalists, however, look to the costs and consequences of economic development, and as such are forced to consider a wider scope of concerns and issues. More variables means greater complexity, and with that comes increased data requirements, fewer tractable solutions, and less clear theories. For example, the well intentioned efforts of environmental economists to bring natural factors within the analytical techniques of economics have failed to produce meaningful results (Economist 1998). The difficulty comes in pricing environmental attributes to fully account for economic activity. For instance, when a tree is chopped down and sold for lumber, the cost someone would be willing to pay to look out over an unblemished forest must also be considered. Such analyses, besides being highly variable and hypothetical, miss more basic questions: For example, what does the tree contribute to the more complex web of environmental processes, and what is the tradeoff between short-term economic gain and long-term environmental health? Such questions cannot be answered by modern economic methods, and it is not clear that simple modifications will allow them to do so in the near future (Cassidy 1996). Despite such methodological difficulties, the reason for economics' influence is clear—the ideas implicit within it continue to shape the lives of people across the globe as capitalism forms the core of the modern social environment. This leads one to question whether other intellectual traditions might better capture the complex essence of the natural environment.

1.2.2 Geopolitics and Power

Those who study politics and markets seldom address topics at the interface between the social and natural environments. When they do, the impact of the environment on economics is far more likely to be the topic than the impact of economics on the environment. Geopolitics and power, in sharp contrast, address precisely these questions by including a wide range of physical factors into the analysis. Thus, geopolitics takes seriously issues of scale, scope, spatial distribution, interconnectedness, and complexity, and by doing so provides more of a perspective or framework for study rather than a well defined area of study. The historical relationship between politics or *international relations* and geopolitics or *geography* is long, derivative, and fraught with difficulty. This section describes the relationship's history as well as its relevant modern forms. The intent is not to present an exhaustive history of geopolitics and power but instead to provide contrast with politics and markets thereby yielding perspective for the subsequent analysis.

The "taproots" of international relations extend not into law or economics but into geography (Renouvin and Duroselle 1964, chap. 1). The connection between these related fields is brought into stark relief when one considers that the very notion of "state" is bound up with territorial control. Once this is acknowledged, it is but a short step to cataloging and comparing the natural resource endowments of different countries and asking how they might affect geopolitics and power. Since international relations at its heart concerns conflict and cooperation, early studies took on a distinctly martial flavor. Climate, topography, hydrography, and soil were all studied from the perspective of how they contributed to national power or influenced the national temperament. Geographers noted that national economic development first occurred in temperate zones, that topography influenced the density of settlements, the mingling of populations, and the boundaries between nations. Rivers determine the development of trade routes, the placement of cities, and the movements of armies. From the topsoil came the food to feed armies, and from the subsoil came the energy and minerals needed to fuel and build industrial societies. Demographic variables were also examined with an eye towards power calculus as soldiers themselves were culled from the general population.

While the importance of geographic factors is readily acknowledged, initially too much explanatory power was placed in them. Early geographic theories explained too much about politics based on too few environmental variables, rendering the resulting theories deterministic. For example, it has been argued that Great Britain's propensity to expand was driven by its being an island nation. In contrast Japan, also an island nation, has a history marked by isolation (Aron 1966, 191). Clearly there is more to a nation's tendency for expansion than whether or not it is an island.

Determinism manifested itself in the writings of geographers throughout the late nineteenth and early twentieth centuries. Friedrich Ratzel sought to renew the foundations of political science by articulating connections between society and the land, between the social and natural environments (Renouvin and Duroselle 1964, 18). His vision was pursued through a set of biological metaphors in which the state was seen as an organism in conflict with others. A nation could increase its available resources and security simply by increasing its size. Sir Halford John Mackinder simplified, distorted, and popularized Ratzel's work with his theory of the heartland—the large, flat area of rivers and grassland in Eastern Europe. Mackinder maintained, "Anyone controlling

Eastern Europe controls the Heartland. Anyone controlling the Heartland controls the World Island. Anyone controlling the World Island controls the world" (Aron 1966, 192). These sweeping assertions—unencumbered by rigorous empirical or causal analysis—aspired to science but amounted to ideology (Aron 1966, 197).

German geographer Karl Haushofer furthered Mackinder's work by making it more explicitly political. Haushofer held that geography drove politics and that through geography, politics could be described in terms of palpable facts and provable laws that would, in turn, drive policy. The policy prescription that has proven most lasting (or infamous) is *Lebensraum* or "living space" (Renouvin and Duroselle 1964, 19). Frontiers, rather than being immutable, were seen as malleable depending on national problems and priorities. Returning to Ratzel, who viewed countries as organisms capable of growth, it is easy to see how National-Socialism perverted this biological metaphor into the more popularized, power-oriented, and aggressive German *geopolitik*. Just how intentional were Haushofer's contributions to German adventurism remains questionable. More certain is that when Haushofer objected to Hitler's planned foray into the Soviet Union, his services were no longer required by the Nazis.

After World War II, geopolitical studies waned for many reasons, not the least of which were its perceived contributions to German aggression. Beyond that, the contributions of nature and natural processes to human society and its politics were so complicated that geopolitics allowed for the expression of profound questions but could offer few answers (Aron 1966, 196). Moreover, rapid technological advances after World War II moved political analysis away from the natural environment, and the Cold War reduced geopolitical considerations to bipolar, global-level conflicts, a stance that plainly discouraged focusing on local complexities. These developments simplified politics in that the details of scale and scope no longer merited serious analytical interest. Recently, geopolitics has again become popular as the "quantitative revolution" of the 1950s and 1960s has provided new analytical techniques for geographers to address complex issues, and the end of the Cold War has allowed small-scale social and natural analyses to be undertaken afresh (Taylor 1993, xi–xii).

The *new geopolitics* describes the "rebirth" of a discipline that addresses, as did the old geopolitics, questions of politics and territory (Ward

1992b).³ If politics is the art of achieving desired goals, then territory constitutes that which is most often desired. This observation brings to mind conflicts between states over borders and the demarcation of national territory. Such state-centered analysis has dominated international relations over the past seventy years. These years have seen the creation of many new states, two world wars, the Cold War, and a host of smaller-scale conflicts including Korea and Vietnam. That the twentieth century has proven a period of violence and natural resource exploitation is to be expected as the modern nation-state asserts a monopoly on violence, war making, and resource extraction. Where the new geopolitics differs is in the scope of its analysis. To argue that interstate conflict concerns only the placement of borders belies the multicausal complexity of the political processes at play. "Throwing the state back out" captures the realization that causal factors besides the state and its attendant actors and policies must be considered when characterizing the processes that compose modern world events.

This change in perspective becomes apparent when contrasting the use of organic thinking. Classical geography used the organism both as the foundation for analogies and the justification for predation: so goes the reasoning, "States are like animals, animals consume other animals, therefore states should consume other states." The New Geopolitics retreats from such syllogisms, recognizing instead that "organic-ness" refers to the ephemerality of states. Although they may seem permanent, states are not immutable fixtures on the world stage—instead they are created, maintained, and abandoned when no longer useful. This view recognizes that states are made up of multiple, diverse elements and that the failure of a single, internal element can prove just as nationally injurious as an invasion from another state. Thus, state failure can occur due to intranational as well as transnational processes.

The new geopolitics therefore constitutes a recognition of multiple relevant factors or *complexity* as state security and survival now require more than simply sufficient military power (Ward 1992b, ix). The Soviet Union was, after all, one of the world's two military superpowers, and in reviewing its demise one begins to appreciate the importance of other, nonmilitary security factors, including democratic representation, economic militarization, and endemic corruption, all of which proved as relevant to state security as military strength or ideological commitment. Because internal institutions like governmental and economic organizations are determined by the sovereign state, the state itself must be judged by its ability to provide and maintain (1) sovereignty for itself, and (2) security for its citizens. These are not as separate as they might first appear; the failure of the Eastern Bloc countries at the end of the Cold War shows that regardless of how powerful or totalitarian the state, the security of citizens directly affects the sovereignty and viability of the state.

Acknowledging and understanding the complexity of geopolitics also counters the determinism of earlier analyses by introducing probabilism or possibilism. The first international relations theory to address such issues is the *ecological triad* of entity, environment, and entity-environment interaction (Sprout and Sprout 1965, 1968, 1971). Although the ecological triad is reminiscent of a Hegelian dialectic—thesis, antithesis, and synthesis—it is more interesting to recall Ratzel's desire to include society and nature in the study of politics. Like Ratzel, the ecological triad asserts that environmental concepts are critical in human affairs. However, the modern version holds that perceptions of the environment lead to decisions, and the environment influences or limits the efficacy of those decisions. In this manner, the environment influences far more than it determines the course of human affairs.

More recently, possibilism informs the theory of "opportunity and willingness" (Most and Starr 1989; Starr 1992), which not only accounts for environmental contributions and uncertainty in war studies but hearkens back to those geographers who argued that nature gives society opportunities, but humanity remains their ultimate master (Febvre 1924; Renouvin and Duroselle 1964, 20). Recall that geography offers a perspective based on spatial distribution and proximity, a perspective that encompasses technology, ideology, religion, and government. The geographic perspective addresses both large-scale, macrolevel questions and small-scale, microlevel ones. Taking proximity seriously means that decision makers place more weight on local information, which forms the basis for more cognitive theories such as bounded rationality (Simon 1983, 1985). Analyzing large and small-scale factors simultaneously brings into play the problems of aggregation, abstraction, and levels of analysis.

Specifically, how does one incorporate both microlevel behavior and macrolevel environment—that is, *context*—into a unified and coherent analysis? This is a hard problem, but the opportunity and willingness framework grants some purchase in studying political behavior and the environmental context that shapes it. Opportunity represents possibilism, the range of plausible options offered by the macrolevel environment. Willingness represents probabilism under choice, that which tends to occur given one's belief system, operational codes, and mental maps (Starr 1992, 5). *Context* thus represents the structure of opportunity and willingness that implies the synthesis of microbehaviors and macroenvironments in both a spatial and temporal sense. The importance of context regards the incentives the environment presents to the decision maker, including the costs, risks, and benefits used in any expected utility calculus. Such incentives are thus at the nexus of combining macro- and microlevel analyses. Context also embodies the notion that applying decision rules depends on time and space—that is, space is heterogeneous and decisions are domain specific—and so it represents a move away from universally applicable rules and absolutism.

Context applies to multiple scales, from the local, microlevels that influence personal action to the geopolitical, macrolevels that constitute the international system. Such macrostructures are analyzed through *world system theory*, of which Braudel (1979), Wallerstein (1979), and Taylor (1993) provide a reasonable and representative sample. Braudel argues that expansion and development of the world economy has been a major, driving force in world history. Wallerstein formalizes Braudel's notions of "core" and "semi-periphery" and "periphery," which captures notions of structure and the ordering of space in studying social systems. "Core" and "periphery" are *horizontal* notions that connote spatial differentiation among geographic units of similar scale. *Vertical* differentiation denotes variation among units of different scales (Taylor 1993, 44): Exemplary typologies include (1) locality, nation-state, and world economy, (2) home, national economy, and world market, and (3) experience, ideology, and reality. Such orderings reveal a nuanced perspective seeking to provide analytical order and understanding in a complex world.

Economics, in contrast, seldom considers why separate areas develop differently, viewing uneven development more as a problem to be corrected than a phenomenon to be understood (Krugman 1991); international relations also tends to view territory simply as an undifferentiated, homogeneous space or *isotropic plain* (O'Loughlin and Anselin 1992). Core and periphery acknowledge the detail and variation that comprise the world as it is rather than as it should be. The basic problem centers on system complexity—entity and environment, probability and willingness, horizontal and vertical differentiation all try to get at some basic, fundamental notion of system. It can be said that all systems, including world systems, exhibit both structure in space and dynamics over time, but the relationship between structure and dynamics is subtle and difficult to comprehend fully. However, the quantitative revolution renders the mysterious interplay between system structure and dynamics increasing accessible.

1.3 The Role of Computation

Economics, insofar as it has brought about today's international system, has been characterized as the social environment's exemplary discipline. The quantitative revolution of the 1950s and 1960s (Taylor 1993, xi–xii) has contributed differentially to economics and geopolitics, rendering traditional economics increasingly problematic (Cassidy 1996), while simultaneously transforming traditional geopolitics into the new geopolitics (Ward 1992a). The driving force behind the quantitative revolution is the computer: recent increases in computational scale have led to changes in analytic kind. This section describes the changes wrought by enhanced computational capability as they affect the social and natural environments.

Computation's fundamental impact concerns the increasing transparency and accessibility of high-dimensional complex systems. That is, far more system components and interactions can be tracked with a computer than with the human mind. Consequently, computer-based techniques have helped emphasize the experimental and inductive at the expense of the theoretical and deductive (McCloskey 1995). Economists traditionally proposed theories and policies supported by highly aggregated statistical studies, which was reasonable given the historical expense of data collection and computation. With the advent of powerful computers, the empirical and statistical aspects of economics have been greatly strengthened. Far more types and greater quantities of data can now be collected and analyzed, but economic theory has not progressed commensurately—links between micro- and macroeconomics remain underdeveloped. However, this divide is narrowing as traditional statistical techniques give way to more powerful simulation techniques. Beyond the aggregated statistical techniques that form the basis of traditional economics are computer simulations that more directly represent the multiple, microeconomic forces that contribute to large-scale, macroeconomic phenomena. So instead of theorizing and postulating how individual sectors might influence the aggregate economy, computer simulations allow for more direct representation of the economy's small-scale *microfeatures* and direct measurement of the economy's large-scale *macrobehavior* (Forrester 1989).

Being able to cross the analytic divide between micro and macroscale phenomena, including microfeatures and macrobehavior, is a fundamentally new and powerful way to analyze complex socioeconomic systems. Without the computer to handle the myriad causal relationships and niggling mathematical details, the relationship between microfeatures and macrobehavior would remain hidden. Nevertheless, computation's contribution is not purely empirical and data related. Using the mathematical language of physics and the natural environment—differential equations—current simulation methods in fact deemphasize the importance of time-series data, emphasizing instead the causal relationships among the low-level elements that compose the microfeatures, which in turn generate the system's high-level macrobehavior (Forrester 1971, 55; Sterman 2000). The key is not to maintain fidelity with empirical data but with the system's underlying causal structure.

Performing such analyses requires a different set of motivations, assumptions, and conceptions as well as a new language that allows for their expression. The first problem is one of microfeatures or system complexity: how to express system structure and the causal connections among disparate system components? Four aspects of complex systems are discussed herein: (1) stock and flow relationships, (2) nonlinearity, (3) feedback, and (4) hyper-connectedness.[4]

With differential equations, not only are the values of variables important, but so too are the rates at which those variables change. That variables change over time is a key insight because it introduces the notion *time, motion, dynamics,* and *behavior* that characterize complex systems, both social and natural. For example, consider water in a bathtub. The water level changes over time because bathtubs have spouts that fill the tub and drains that empty it. In the argot of complex system simulation, the tub is a *stock* and the spout and drain represent *flows* (HPS 1997). Stocks thus represent values or levels, and flows allow the level of a stock to vary over time, making the model dynamic. Flow rates may be based on causal relationships with other system values—for example, an increase in *A* leads to an increase in *B*—and this relationship may be nonlinear as depicted in figure 1.1. Nonlinearity makes the mathematics more complicated, but it pays dividends by allowing systems to be represented more accurately.

As the web of causal relationships thickens among system components, they tend to "feed back" on themselves as shown in figure 1.2. That is, as *A* influences *B*, *B* can in turn influence *A*, forming a *feedback loop*. Multiple, interconnected feedback loops such as those among *A, B,* and *C* and can be represented using differential equations and simulated. Feedback, like nonlinearity, makes the system both more complex and realistic.[5] Finally, large simulations include many causal connections that address

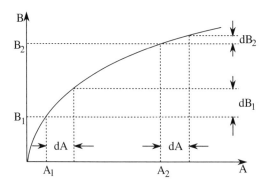

Figure 1.1
Nonlinear Causal Relationship. In this nonlinear causal relationship, A causes B, or more precisely, cause A1 leads to effect B1 and A2 to B2. The nonlinearity of the system is demonstrated by the two equal *changes* or dAs applied at A1 and A2, which result in two very different dBs. A linear system would yield dBs of equal size.

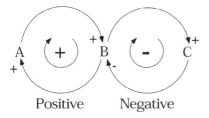

Figure 1.2
Feedback Loops. There are four causal relationships in this system: (1) A causes B, (2) B causes A, (3) B causes C, and (4) C causes A. The arrows denote causality and *feed back* on themselves to form loops. The signs next to the arrows denote polarity. A plus (+) sign denotes that cause and effect vary in the same direction—for example, as the cause gets bigger, so does the effect; a negative (−) sign denotes change in the opposite direction—as the cause gets bigger, the effect gets smaller. These individual causal relationships together form feedbacks, which also have polarity as denoted by the circular arrows in the center of the loops. A positive arrow denotes a "self-reinforcing" feedback that grows; a negative arrow denotes a "goal-seeking" or "balancing" feedback that seeks a constant level.

the *hyper-connectedness* of complex systems. It has long been recognized that in the real world everything is connected to everything else; modeling this is impossible. Creating good models requires a certain level of conceptual pruning, including the important and germane and ignoring the inconsequential and superfluous. Combining stocks and flows with nonlinear and feedback relationships creates a system of differential equations that enables the analysis of system driven, dynamic disequilibrium models as opposed to the data driven, static equilibrium models favored by economists. Doing so allows researchers to analyze old questions from fresh perspectives, which leads to new insights and more effective policies.

Issues of system representation have consequences beyond their mathematical details. Multiloop, nonlinear feedback models allow for renewed investigations into several classes of economic problems. For example, consider the modeling of human cognition, which in microeconomics has overestimated human cognitive processing capability and the availability of high quality information. Using simulation, human behavior can be modeled in a more *boundedly rational* manner observed by experimental psychologists (Morecroft 1983; Simon 1985; Forrester 1989, 9). Specifically, "real-world" decision makers tend base their actions on few and certain local data, which squares with the geographic perspective that nearby influences be weighted more heavily than those further away. Additionally, simulation methodology allows for a more accurate market clearing model. Traditional economics bases the behavior of markets on price while multiloop, nonlinear feedback models allow for the modeling of delivery delay, the time a customer waits before receiving a product. Delivery delay is not included in standard economic analyses, but experience shows that the time it takes to get a product is just as important as price in determining actual market behavior (Forrester 1989, 8). Differential equation models thus allow one to control more precisely the number of factors being analyzed: with respect to bounded rationality, the number of relevant factors is pared down, and with respect to market clearing, the number of relevant factors is increased. In both cases though, fresh insights are obtained because the resulting model more closely represents reality.

Multiloop, nonlinear feedback models also allow for the more complete representation of a system's macrobehavior. The potential complexity of a model's microfeatures and its relative freedom from data allows for causal factors to be made *endogenous* or brought within the model. This then creates a different flavor of modeling—whereas traditional statistical models feature few causal connections and copious amounts of data, differential equation models feature richer causal connections with less stringent data requirements. In this manner, initially simple models can be made more complex as the modeler learns more about the problem, especially as one discovers that initial problem conceptualizations prove inadequate in capturing the actual behavior of a complex system. In other words, factors initially thought to be *exogenous* or external to the problem at hand can be "brought inside" simply by expanding the model. This process implies analytical iteration through the building of microfeatures and the testing of their macrobehaviors that can take on the character of informed trial and error, play, scenario analysis, or even policy evaluation. The key aspect of this iterative process is that learning takes place as the modeler changes the system's microfeatures and then observes how those changes affect the macrobehavior.

To summarize, increases in computational capability have led to commensurate increases in model complexity. This bears on the tension between the social and natural environments because, even though both environments are acknowledged to be complex systems, the social sciences have historically been analytically simple and the natural sciences historically complex. As regards this study, the increased capability to represent and analyze complex systems will be used to establish causal connections between the social and natural environments and to encompass and reconcile both sides of the the pro- and antiglobalization debate.

1.4 Meeting the Challenges

This chapter began by reflecting on Seattle's 1999 WTO riots. While discovering the exact cause is practically impossible, it is reasonable to say that the divergent views held by economists and environmentalists regarding the relationship between trade and the environment is a significant source of friction. Two divergent literatures reveal that economists think trade helps the environment and environmentalists think trade hurts it. Rather than pursuing the question through ongoing protests, this study seeks an analytic answer to the impasse, one based in systemic representations that connect trade and the environment. Economics as a field and the international economy as a system rest on a foundation of fairly simple theories because high quality computation has only been available for the last few decades. Since high quality computers are now available, it makes sense to employ them in understanding globalization and trade.

The new analytical methods made possible by enhanced computation allow linking of the social and natural environments through multiloop, nonlinear, feedback models. Such models achieve their analytic strength by accurately representing the causal connections that comprise a complex system. Thus, a system's microfeatures can be accurately represented, and its resulting macrobehavior can be tested and analyzed. As such, no longer should the analytical exercise be considered solely the interplay between theory and data but the interplay among data, theory, and computation (Rowell 1995), with the result that problems of greater complexity can now be addressed analytically. In acknowledging the complexity of the social environment and the success of economics, global environmental degradation can be thought of as a byproduct of analytic simplification, also called the intellectual's search for the single cause (Woo 1990). These new, simulation-based models turn this search on its head, transforming it instead into the search for multiple causes and aggregated effects. Analyses are made more complicated rather than less, and this complexity aids in the comprehension of system features and the behaviors that derive from multiple, interacting components.

How can one harness these insights in studying the political and economic contributions to global environmental degradation? Establishing correlations among high-level demographic and environmental variables is not enough. The goal is to drill down into the local political and economic details to understand how personal-level decisions and actions affect the global environment—how the social environment affects the natural. This requires a way of analyzing how low-level political and economic features aggregate up to form high-level global environmental behaviors. Standing in the way are the multiple levels of analysis between the local and global scales. The methodological advances described herein provide new ways of addressing such complexities, and establishing a better understanding of global environmental degradation provides an appropriate venue for the endeavor.

Despite their explanatory power and mathematical sophistication, advanced methodologies do not solve problems by themselves. They require focus and direction. Without it, quantitative methods can confuse as much as they clarify. Better theories are therefore required to guide and inform their application, as it is understood that theory comes before experiment and analysis. This study draws from international relations theory generally and lateral pressure theory specifically to organize ideas and think clearly about global environmental degradation. Informed by the intuitions developed here, the next chapter specifically addresses the task of theory development.

2
Lateral Pressure Theory

When united, globalization, trade, and the environment form a large-scale and complex system. With so many of the world's regions, countries, cultures, economic systems, and ecosystems connected in so many different ways, it is easy to be overwhelmed by the multiple, interconnected factors that compose the *environmental problématique*. Forests, croplands, rangelands, rivers, groundwater, wetlands, oceans, coral reefs, islands, mountains, deserts, and a myriad of other ecosystems have all been impacted by human activity with each featuring its own attendant, evocative, and compelling photographs chronicling the destruction (Goldsmith et al. 1990). While addressing individual ecosystems is an understandable and important undertaking, such studies miss the inherent interconnectedness of the international system and of global environmental degradation.

Synthesizing disparate studies into a coherent whole might achieve a global perspective, but doing so is hard. First, the data, methods, and results from the studies must be compatible. Second, even if they do fit together, their results still need to be pared down into manageable chunks, and that requires a defensible criterion for idea inclusion, an intellectual razor for result separation, or a conceptual lens for research focus. Third, the studies, when combined, must form a larger argument that supports or contradicts some previously held view of the world. Such a view is called a *theory*, which is simply a supposition, speculation, or conjecture that focuses research effort and guides experimentation. Thus, theory comes before research rather like blueprints before building. Experimental results will either agree or disagree with the motivating, organizing theory, and if they disagree, then the theory—in some way, shape, or form—is wrong.[6]

An organizing theory is, therefore, needed to guide the analysis of global environmental degradation. Theory development is made possible by questions or puzzles that provide fundamental insights that can then be used to formulate theories that direct analysis. The first inkling of such a puzzle comes in considering the many different types of environmental degradation that are present across the globe. It would be unusual for so many of the world's ecosystems to be affected simultaneously—that is, throughout the twentieth century, and especially its latter half—simply by chance. Thus, there must be some synchronizing process that effectively coordinates more localized environmental degradation. Panayotou (1992, 317) clarifies this puzzle by observing,

> Environmental degradation is a more common and pervasive problem than rapid inflation, excessive foreign debt, or economic stagnation. Rapid deforestation, watershed degradation, loss of biological diversity, fuelwood and water shortages, water contamination, excessive soil erosion, land degeneration, overgrazing and overfishing, air pollution and urban congestion are as common to Asia as they are to Africa and Latin America. It is striking that rapidly growing Southeast Asia has similar environmental problems as stagnating sub-Saharan Africa or heavily indebted Latin America.

Panayotou asks, why are political and economic problems are so varied but environmental problems so similar? In so doing, he contrasts the *social environment*—politics and economics—with the *natural environment*—two different realms with two different sets of problems, interests, and traditions (Choucri 1993b; Wilson 1998, 10). Moreover, the scale of analysis is inherently global—ranging from Southeast Asia to Africa to Latin America—and yet the environmental degradation experienced in these distinct locations appears to be independent of regional variation. Finally, although human activity has long caused localized environmental degradation (Whitmore et al. 1990; Ponting 1991), never before have regions as geographically separate as Southeast Asia, Africa, and Latin America been simultaneously subjected to such all-encompassing and similar—that is, *global*—forms of environmental degradation. Panayotou's observation provides the fundamental puzzle for this study by (1) comparing the social environment with the natural, (2) contrasting the global scale with the local, and (3) raising questions of synchronization, connection, and linkage across environments and scales.

The social and natural environments, until recently regarded as analytically separate, are again recognized as inexorably linked through the specter of global environmental degradation. However, it remains to be seen just how the social environment, both politics and economics, fits into this burgeoning constellation of concepts. For this study, *Lateral Pressure Theory* is forwarded to link the social and natural environments, and in so doing it addresses the phenomenon of interregional synchronization through globalization generally and trade specifically. In undertaking the task of theory development, one need not start from scratch. Much work has already been done in this area, and it is reviewed before addressing and extending lateral pressure. In so doing, a theoretical foundation will be established that directs, organizes, and focuses the analytical efforts that follow.

This discussion is divided into six sections. First, traditional theory is revisited to establish relationships between the social and natural environments based on philosophical underpinnings. Second, international relations (IR) theory is examined to understand how current international-scale studies are ordered. This discussion takes place in terms of IR's traditional topic, interstate conflict or war. Third, lateral pressure, this study's organizing theory, is examined to understand how it differs from traditional IR theory. Specifically, whereas traditional IR theory is more grounded in economics and the social sciences, lateral pressure derives its insights and analytic rigor from the natural sciences. Fourth, lateral pressure is extended from the traditional IR topic, war, so that it may address questions encompassing economics and the environment. Fifth, analytic methodologies that support lateral pressure theory are identified and discussed in this chapter and used more intensively in subsequent chapters. Finally, the key questions that motivate this chapter—the social and natural environments, the theories that span them, and the methods that link them—are revisited.

2.1 The Classics

In addressing the social and natural environments, one discovers that the topic is as old as philosophy itself. Plato's *Republic* conceives of community, or the *polis*, as a middle ground between nature and the individual, between the *cosmos* and the *soul*. This observation is rooted in physiological necessity as people, unlike animals, are wholly dependent on those around them for many years until they are mature enough to make their way in the world (Gaylin 1989). At the most basic level, the community or polis makes it possible for the individual to live in nature—for a soul to exist in the cosmos. This is philosophy at its best, giving form to an idea so that it may later be theoretically developed and empirically supported. Like the atomic theory of Democritus or the evolutionary ideas of Empedocles (Sagan 1980, 179–180), Plato's articulation of the cosmos, polis, and soul is not the final word on the issue as there are many more details to be thought through. Science is not just about finding the right answer, it demands to know how one arrives at the answer. Thus Plato's observation constitutes not a conclusion but an intellectual point of departure.

Cosmos conveys the notion of an orderly, harmonious, and systematic universe, an order traditionally understood through the study of regularities or laws. Laws come about through the power of nature or the authority of the state (Hart 1961, chap. 9). A natural law, like the law of gravity, is a regularity that derives from the primacy of physics. A social or state law, like a speed limit, is a behavioral regularity that derives from a combination of general consensus and the executive enforcement of sanctions. These two types of law, the natural and social, combine when a nation-state collectively determines how it will use its national resource endowments. As a nation's population and technological capacity increases, so does its natural resource use. Consequently natural resources can become strained or threatened through depletion or pollution, and so legal institutions are called upon to regulate these resources to ensure their long-term viability (Ostrom 1990, chap. 4). Thus, there exists a closely coupled relationship between natural and state law, and it should be expected that environmentalists, in seeking to protect the natural environment, will do so by working through the state's legal institutions.

As clear as the relationship is between natural and state law, there are several difficulties that complicate the matter. First, environmental degradation can take place at many different scales, from the very small to the very large. Some environmental problems are so large that they range across national borders. Given such a situation, which state's legal system should address the matter? One might respond that all concerned states should be included in diplomatic negotiations, but consider the options should one state be difficult and choose not to cooperate; what then? Further negotiations, economic incentives or sanctions, and military intervention might come to mind, but appealing to a stronger authority is not a truly viable option because there exists no institution above the state that can effectively compel the state. This makes international law fundamentally different from law within a state because no higher authority exists (Hart 1961, chap. 10). Consequently, global environmental degradation, because of its global scale, forces environmentalists to operate in the realm of international law and treaty formation (Weiss 1992; Hass and Sundgren 1993).

A second complicating factor is the considerable gray area between natural and state laws. All manner of activities are performed without state enforcement or centralized direction either because of tradition, they've always been done this way, or because such behaviors spontaneously manifest themselves. One might consider the fashions of American youths that tend to exhibit high levels of nonconformist conformity. Decentralized social regularities can also take place within other communities. Within the community of nations, social regularities are called *regimes*, the implicit and explicit principles, norms, rules, and procedures that guide international behavior (Krasner 1982, 1983). Here regularities manifest themselves not due to a centralized legal system but through a process of decentralized consensus and bargaining. International regimes, while not explicitly dictated and directed, can be shaped and influenced by international institutions that serve as a repository and memory for the aforementioned principles, norms, rules, and procedures. Moreover, the regime concept can be explicitly applied to securing cooperation among nations in addressing environmental concerns (Young 1989, 1993).

Excessive focus on the classic, social concerns of law, regimes, and institutions can belie that which is of major concern—the health of the natural environment. A "second wave" of research recognizes this by asking whether or not international environmental regimes indeed help to improve the environmental problems they are intended to correct (Soroos 1994, 303). This perspective acknowledges the natural environment as analytically coequal with the social, an unusual premise for an essentially political study. For example, Haas, Keohane, and Levy (1993) present a series of case studies ranging from ozone depletion to acid rain to fishery management. Another well developed case is that of intentional oil discharge from ships at sea (Mitchell 1993, 1994). In each case, a particular environmental problem is isolated and an historical account of regime development and its effectiveness is presented.

Case studies provide valuable, initial glimpses into specific environmental problems, and they point the way for further theoretical work that can knit together seemingly disparate issue-areas by identifying and articulating similar underlying causal structures. Choucri (1993a) provides an initial attempt at formulating such a "third wave" theory. Choucri's work is, on the whole, more abstract than those previously mentioned as it seeks to articulate a more comprehensive theoretical framework by considering a global system comprised of two large, complex, and interconnected systems, the social and natural environments (cf. section 1.2). The task of connecting these two complex intellectual traditions and systems, as shown in figure 2.1, is a "linkage challenge" (Choucri 1993b, 3–4), the solution of which would prove an important contribution to the emerging field of environmental politics. This challenge is addressed by specifying causal relationships between the social and natural environments that answer the following sorts of questions: What links the two environments? Where do these linkages apply? How should the linkages be expressed and analyzed? The improved understanding of global environmental degradation requires formulating, articulating, and answering such questions. Before developing a theory that does so, current IR theory is reviewed so that theoretical and analytical extensions can be made from an established analytical baseline.

Figure 2.1
The Linkage Challenge. The *linkage challenge* entails the identification and formulation of causal relationships that span the social and natural environments. The question itself implies that the social environment affects the natural, and the natural affects the social. The challenge lies in connecting the social and natural environments within a single, coherent analytical structure (Choucri 1993b, 3–4).

2.2 Clarifying Concepts and Contentions

Before delving into the intricacies of lateral pressure, a few relevant works are reviewed and critiqued so that lateral pressure may be understood with respect to international relations theory in general. This initial discussion is not intended to be an exhaustive review of the international relations literature; many relevant works worthy of prolonged study will be omitted. The works discussed are intended only to set the stage for the more focused lateral pressure discussion to follow.

At its heart international relations involves the study of war, an endeavor that implies both conflict and cooperation. There are many ways to study war and, as has already been noted, intellectual progress tends to begin with philosophical intimations and end with theoretical and empirical development. Kenneth N. Waltz epitomizes this progression as he starts with a close reading of political philosophy and ends with a deductive theory of international politics. His journey consists of two major steps, from philosophy to static structure (Waltz 1959), and from static structure to dynamic theory (Waltz 1979). Waltz (1959) begins with the following question: What are the major causes of war? In his reading of political philosophy, Waltz pares war's causes to three images:[7] the individual, the state, and the international system (Waltz 1959, 56). The first image, the individual, is included because war is a fundamental consequence of human nature and behavior. Moreover, the individual is the only sentient image of the three. The second image, the state, represents a collection of individuals working together under a common government, only sometimes towards war. The third image, the international system, is marked by state interaction under anarchy. That is, whereas individuals can appeal to a higher authority within a state, states cannot appeal to a higher authority within the international system.

From this study emerges a social structure of scale, a three-level logic, that orders the study of war. The structure is neither crystalline nor compartmentalized, neither static nor separate. Reciprocal interactions among images are explicitly recognized. Individuals influence the state, and the state influences individuals; states influence the international system, and the international system influences states. Any plan for peace must account not only for such interactions but also for the inherent particularities and exigencies of each image, especially those of the international system. Specifically, the anarchy of the international system guarantees that force will continue to be used because there exists no other way consistently and reliably to resolve the inevitable disputes that arise among rational, self–interested states (Waltz 1959, 238).

This conclusion can be interpreted in several ways. First, the recognition of force constitutes an affirmation of realist, power-based politics (Morgenthau 1985 [1948]). Second, the recognition of anarchy constitutes an implicit rejection of the idealism that led to collective security and the League of Nations (Carr 1964 [1939]). Third, and most interestingly, what emerges is a fundamental extension of politics. Normal politics, the type that people encounter most frequently, consists of activities ordered within a commonly accepted social framework, whether that framework is democratic or authoritarian. The enduring promise of world government can be understood as the desire to impose a familiar, state-centered social order onto the international system. The third image holds this is impossible. This is not to say that social interactions under anarchy are chaotic and unstructured (Axelrod 1984). The third image instead represents a fundamental movement away from the "inside-out," state-centered pattern of thinking that prevailed in international politics towards a more "outside-in," systemic conceptualization that acknowledges more decentralized, anarchic, and emergent behaviors and influences (Waltz 1986, 321).

Defining a system's structure is just one aspect of understanding. Another regards how the system's separate and dynamic components interact and behave over time. Understanding a system's dynamics is more complicated than understanding its structure because dynamics presuppose structure.[8] Waltz (1979, 5–6) steps away from a structural, associative, and law-based understanding of international politics towards one that is more dynamic, explanatory, and theoretical. Waltz's theory development is explicitly deductive as induction implies, "that we can understand phenomena before the means for their explanation are contrived" (Waltz 1979, 7). Theory construction for Waltz is not labor intensive; instead, it requires a flash of intuition, a creative spark. Theory construction is an exercise in "emergence," with the whole being greater than the sum of its parts. It is not an exercise in reduction, detail, or data gathering. However, such activities can be useful in theory construction if coupled with simplification, abstraction, aggregation, and idealization because they aid theory visualization. Only when a theory exists can the process of explaining occurrences and recorded associations begin (Waltz 1979, 9).

The theory proposed by Waltz (1979) centers on notions of "system," and much of the discussion is spent teasing out system's implications and consequences. The system approach is defined in relation to the analytic approach. Where the analytic approach considers component interactions in reduced dimensionality subsets, the system approach acknowledges the full complexity of multiple, simultaneous component interactions (Waltz 1979, 39). The system sensibility, when applied to international politics, reveals that the international system cannot be understood merely by "looking inside" of states (Waltz 1979, 65). This is not to say that image levels are insulated from one another, that nations are insulated from the international system. They are not. There are, in fact, intense interactions among image levels, and it is through such interactions that the comparatively subtle and abstract level of analysis, the international system, reveals itself. States, while they are indeed sovereign, continually find that their policies do not always turn out as planned just as an individual's personal plans can go awry. Such *unintended consequences* are the result of complex systemic constraints, of persistent environmental pressures placed on the nation or individual (Waltz 1979, 65). Even though the international system is inherently anarchic, the workings of the international system are able to exert constraints and pressures that shape the states within it. The enduring character of the international system speaks to the ubiquity and potency of these shaping forces.

Noting the consequences of a process, however, is not the same as explaining or articulating it. The question then becomes, how to conceive of order without a centralized, ordering entity? Here, Waltz turns to microeconomics in arguing that order arises from multiple and spontaneous self-interested and *rational* actions (Waltz 1979, 89). Rational decisions require a utility calculus and variables on which to calculate: Waltz chooses power, the standard realist measure. Power is estimated by comparing the capabilities of states, a definition that accounts for system in that capabilities may be a function of individual states, but their distribution is a function of system (Waltz 1979, 98). Separate states within the international system are then treated as companies within a market, albeit with different preference structures. In a market economy, companies seek efficiency and it is understood that some will not survive, while in a state system, survival is valued over efficiency (Waltz 1979, 137). Extending from this initial premise, market equilibrium corresponds to the balance of power, increasing market share to imperialism, market stability to peace, and market adjustment to war.

Such sweeping pronouncements were bound to provoke a reaction, and they did. Keohane (1986a) presents a range of criticisms from a range of authors,[9] not all of whose views are represented here. Instead, the responses are interpreted and grouped into two categories, system and power criticisms, while recognizing the risk of oversimplifying the authors' original intent. These two criticism categories are then reviewed with respect to complex systems theory. The system-level criticisms are essentially structural in arguing that life generally and international politics specifically is more complex, varied, and multifaceted than Waltz's analysis makes them out to be. This argument holds that Waltz's three image structure does not adequately account for the workings of the international political economy or the internal workings of states (Keohane 1986c, 158–159). However, the argument also applies if one accepts Waltz's structure. The lack of demonstrated interaction among image levels, especially between the state and international system, calls for advances in domestic politics, decision making, and information processing so that the gaps between analytic levels may be bridged (Keohane 1986c, 191–192). Waltz's proffered structure is too crystalline and brittle, and his analysis too focused on the international system, to explain any meaningful change. What remains is an adequate conceptual framework for question and hypothesis formation (Keohane 1986c, 159).

These criticisms of system carry over into the discussion of power. First, Waltz's use of power is characterized as overly aggregated (Keohane 1986c, 191). That is, the national endowments and capabilities that drive power's value are more varied and contextual than can be represented in a single, comprehensive variable. This has behavioral consequences as calculations based on aggregate measures of power yield notoriously inaccurate results (Keohane 1986b, 11). This point is expanded by Ruggie (1986) who argues that Waltz's theory is unable to account for history in that it does not generate plausible explanations or scenarios. Criticism is also directed at the ambiguity of balance of power, the lack of structural detail, and the inability of the theory to incorporate historical facts into its explanations, all of which lead to the conclusion that a fundamental dimension of change is missing from Waltz's work (Ruggie 1986, 142). Keohane (1986b, 12) suggests that Waltz could benefit from incorporating the bounded rationality work of Herbert A. Simon, but operationalizing these findings, as important as they are, remains an ongoing opportunity for international relations and political science.[10]

Let us pause for a moment to review what has been said regarding neorealism while bearing in mind the following observation: "I would suggest that if we were to look for a core to political science, it would have the character of a relatively small number of questions that will not go away" (Cropsey 1990, 43).

The neorealism debate seems to have bumped up against the enduring question of model complexity. To argue that power is overly aggregated is, in some sense, uninteresting because realists know perfectly well that power is composed of many elements.[11] Similarly, to argue Waltz's theory is overly spare and that it does not account for change is off the mark. Waltz (1979, 10, 210) explicitly states that simplicity and the explanation of change are goals of his theory. Just because goals are stated does not mean they are achieved, which prompts Ruggie (1986, 141) to ask whether Waltz succeeds on his own terms. The answer would have to be no, but it is instructive to analyze why this is so.

A conflation in the debate provides an insight into the theory's shortcomings. Arguments blend confusingly when Keohane suggests Waltz look to Simon's bounded rationality, while Waltz explicitly relies on unintended consequences to reveal system's subtle presence. As it happens, bounded rationality and unintended consequences are different sides of the same complexity coin (Simon 1983, chap. 3). That is, unintended consequences occur precisely because human rationality is bounded and the world is a complex place. In a sense, Keohane and Waltz both argue the same point, that politics is essentially systemic. The only difference is in the phrasing. Keohane articulates system from the individual's point of view while Waltz articulates from the point of view of the environment.[12] Thus, it is always possible to argue that a theory or model is flawed because the real world's structures and dynamics are just too complex to capture with perfect fidelity. That said, some theories and models do a better job than others. While Waltz's theory provides insight into the international system's complexity and nuance, the shortcomings of his modeling strategy—microeconomics-based rational choice—leaves room for improvement. Specifically, rational choice forces the over aggregation of power, the inability to incorporate historical facts, and an overly static analysis. On the positive side, rational choice provides the means to analyze system structure without a centralized, ordering institution, as in the anarchic international system. Nevertheless, a more complex systems-based operationalization might better articulate Waltz's intuitions. This is hard to do, and many international relations scholars have indeed embarked on research programs intended to improve the theoretical linkages between the state and international system (Putnam 1988; Midlarsky 1989; Keohane 1986c, 191–192). The next section focuses on one such effort, lateral pressure.

2.3 Lateral Pressure as Linkage Theory

Lateral pressure theory, at its heart, constitutes an attempt to link national and international level political processes in a dynamic, behavioral fashion. Lateral pressure's conceptual framework for addressing conflict is guided by just these concerns (Choucri and North 1975, 14):

1. domestic growth and the external expansion of interests
2. international competition for resources, markets, and arms superiority
3. the dynamics of crisis.

This conceptual framework is presented in figure 2.2 and is generally applicable to international relations theory. Here we see that lateral pressure addresses the problem originally posed by Waltz (1959): How best to analyze and account for the abstract anarchy of international politics, the more familiar political processes at play within the nation-state, as well as the individuals that drive such processes? Whereas the analysis of Waltz (1979) is economic and parsimonious, Choucri and North (1975) is instead "organic" in recognizing systems as coherent, holistic collections of independent, interconnected and coordinated components. The problem is that Waltz (1979) acknowledges the need for such recognition, but the operationalization of his theory fails on its own terms,

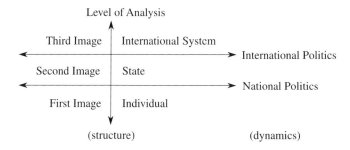

Figure 2.2
International Relations on War. The three images—man, state, and the international system—provide a structure for the study of war (Waltz 1959). Politics is viewed as dynamics within the structure of the system: individuals competing under state-enforced rules constitute national politics, and states striving under the anarchy of the international system constitute international politics.

namely the systemic analysis of a complex and dynamic international system. The test for lateral pressure is essentially methodological and meant to improve on the analysis of the international system's structure and dynamics. The rest of this section is spent describing lateral pressure in light of the criticisms of Waltz (1979), and in so doing discover why lateral pressure might be a good theory to apply in the study of global environmental degradation.

Choucri and North (1975) locate the root cause of war in the processes of national growth that lead to *expansion, competition, rivalry,* and *conflict*. This statement contains elements that tie together international politics, national politics, and dynamics. First, national processes of growth and expansion are explicitly acknowledged. More hidden are their implications for international politics. As nations grow and expand, they come into contact with other nations, growing or otherwise, and the potential for war increases. Competition, rivalry, and conflict all imply multiple nations interacting within the international system. Additionally, expansion bears on the primary geographic variable, space. Expansion denotes a growing national influence, which can cause concern and even consternation in other nations. Finally, *growth* denotes an increase over time within the international system; this does not happen in a vacuum. Growth in one variable tends to affect others. For example, as a national population grows, so too does its need for resources.

The mechanisms for national growth are located within lateral pressure's three "master variables"—population, technology, and resource access—as opposed to the single aggregated variable, power. The three master variables are not offered as inviolate measures. Instead, it is understood that each can be broken down or further disaggregated. Population implies the full range of demographic measures. Technology implies the full range of physical and institutional knowledge as well as skills and infrastructures, while resources imply the full range of natural endowments including land, water, food, fuel, and minerals (Choucri and North 1989, 289–290). Lateral pressure acknowledges that human society is critically dependent on the physical environment and that humanity's mutual dependence on the physical environment provides certain shared experiences and contexts. This enlarged understanding bespeaks a larger commitment to address all levels of analysis—those above the international system, those below the individual, as well as intermediate, intervening levels—all of which makes lateral pressure extensible to issue areas other than war.

National differences in population, technology, and resources correspond to differences in national capabilities and constraints, just as power does for realism. Lateral pressure expresses this notion through a national *profile* (Choucri and North 1989, 304–307). Profile differences correspond to differences in natural endowments and development paths. Thus, profile allows a large amount of data to be incorporated into a reduced number of dimensions, a set of variables more easily analyzed. However, this raises the following question: Does not the singular realist variable, power, do the same thing? While both are indeed exercises in abstraction, the answer is no. The use of the single variable, power, allows for the application of microeconomic techniques including utility maximization and rational choice. The process of grouping countries described by multiple variables into profiles is more aligned with the computer science subfield of pattern matching (Margolis 1987) and classification (Bishop 1995). These latter lines of research correspond much more closely to Simon's work on bounded rationality[13] as previously mentioned by Keohane (1986b). Such techniques bring the power to detect patterns within clusters of multiple variables. The main point is that the national profile technique is not limited to the permutations and combinations of population, technology, and resources—the technique is extensible across a wide range of variable clusters and data patterns.

Verbal descriptions, however, account for only part of the story. The true proof of a concept derives from its formal operationalization. Here, lateral pressure's ability to account for change is addressed with an eye towards verifying that its formal representation matches its verbal description. Methodologically, lateral pressure is less concerned with specific events than with general trends and processes (Choucri and North 1989, 315). Consequently, Choucri and North (1975) address the relatively long-term processes of national economics, colony acquisition, and arms races rather than the short-term processes of warfare and diplomacy. This is done by estimating a set of simultaneous equations using two stage least squares (2SLS) techniques as simultaneous equation models capture the interconnections, dependencies, and constraints within a system. The model creation process breaks down into three stages (Choucri and North 1989, 315): (1) specify the components of the model, both the internal and external variables, (2) decompose the system providing description and rationale for the decisions, and (3) identify a measure or indicator for each dependent variable.

Choucri and North (1975) apply this technique to European international politics from 1870 to 1914, up to the start of World War I. Lateral pressure, the process of national expansion, is expressed as the acquisition of colonies by Great Britain and its primary competitor, Germany. Specifically, domestic causes of colony acquisition are compared against international, systemic causes. Domestic causes are modeled as functions of population density, technology as measured by national income per capita, and access to resources as measured by trade per capita (Choucri and North 1989, 317). Military expenditures are used as an indicator of international-level arms competition. The results from the 2SLS model indicate that British colony acquisition was a function of both domestic and international processes, but arms expenditures were primarily determined by domestic considerations until just before the war (Choucri and North 1989, 319–320). Germany's spending patterns were, in comparison to Britain's, driven primarily by domestic variables.

Choucri, North, and Yamakage (1992) apply lateral pressure theory to pre–World War II Japan. This study's increased methodological sophistication allows for counterfactual analysis and a more comprehensive model (Choucri and North 1989, 319, 322).[14] Colonial expansion proves an important determinant for prewar Japan as well. From 1915 to 1941,

Japan's colonial expansion was predicted solely by its military expenditures, whereas colonial expansion before 1915 was driven by the more domestic need for raw materials. This transition indicates a shift from lateral pressure driven by domestic concerns before 1915 to lateral pressure driven by international-level security concerns after 1915. It was during this period that the United States and the Soviet Union sought to halt further Japanese expansion. The Japanese policy of acquiring colonies to fund its military and increase its security ultimately failed because the additional colonies only served to further increase Japan's military costs (Choucri, North, and Yamakage 1992). When Japan could no longer keep the cycle going, war resulted.

Although lateral pressure reveals much about war, this section began by asking why a theory originally formulated to explain war should be extended towards the study of global environmental degradation. This section has shown that lateral pressure is less a subject area than a systems-based sensibility. More specifically, lateral pressure offers important substantive and methodological advantages over other international relations theories. First, at the outset of this chapter it was argued that global economic expansion since World War II has had significant negative environmental consequences. Lateral pressure is well poised to address the expansion of the international system, and it is explicitly extensible to other issue areas as well. Second, much work has been done on global scale systems within international relations, and much has been learned about systems concepts. Chapter 1 shows that natural systems tend to be more complex and tightly interconnected than can be addressed by standard economic analysis techniques. Thus, the system-based methodologies employed by lateral pressure appear well suited to explore the issues surrounding global environmental degradation.

2.4 Lateral Pressure Extended

Having shown that lateral pressure theory is relevant and extensible, this section formulates the strategy for lateral pressure's application to global environmental degradation. This is done first by formulating a conceptual framework to structure the discussion as was done with the study of war. In so doing, environmental politics is differentiated from national and international politics. This speaks to the continuing question of linking the social and natural environments and of connecting the economic and political with the environmental. Second, this linkage has already been established at the local scale by Ostrom (1990) who develops a more nuanced understanding of the relationships among individuals, political institutions, and local natural resources from a philosophically and analytically informed perspective.[15] Third, this local-scale sensitivity is expanded to the global, systemically connected aspects of global environmental degradation. Lateral pressure is extended with the intention of capturing the international-level, systemic interconnections that contribute to global environmental degradation.

The environmental framework is created by extending the image structure of Waltz (1959). North (1990) develops a fourth image, the *global system,* that distinguishes the social and natural environments just as Waltz (1959) distinguishes the state and international system. The fourth image thesis subsumes the previous three—the individual, state, and international system—and in so doing explicitly links the individual to the global system through the state and international system as shown in figure 2.3. Consequently, the global system (i.e., the natural and physical

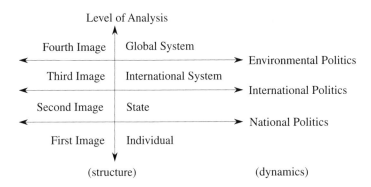

Figure 2.3
International Relations on the Environment. North (1990) adds a fourth image—the natural environment or *global system*—that provides a structure for environmental studies. Politics is viewed as dynamics within the system structure: the international system, a social structure composed of nations competing under anarchy, expanding and affecting the global system constitutes environmental politics.

environment or Earth) shapes and constrains the international system, state, and individual (or social environment) just as the international system shapes and constrains the state and individual. The addition of the fourth image introduces yet another level of abstraction and leads to the same questions of "outside-in-ness" versus "inside-out-ness" that was found in the previous section. Such questions are not addressed further as essentially the same impasse obtains.

Instead, attention is turned to other questions both more pertinent and nascent. Specifically, what is so new about environmental degradation that warrants such attention? Humanity has affected its environment, for better or worse, since time immemorial (Whitmore et al. 1990). Why now? What changed? The difference—and it is a recent difference—is that humanity's aggregate actions now threaten the natural environment in toto. The difference is in the systematization, globalization, and reach of human activity. Essentially, the very scale of human activity has grown too large with respect to the natural environment, so the continued growth of the international system now threatens the viability of the natural environment or global system (Daly and Cobb 1994, 2). The "globalness" of the fourth image therefore implies something intrinsic and important about the natural environment, namely that it too is interconnected, systemic, and holistic. The subsequent analysis must be judged by the extent to which the closely coupled interactions between these two systems, the international and the global, can be captured and explained. The dynamics at this interface between the international and global systems constitute the core concern of environmental politics.

A second question regards the ultimate aim or goal of environmental politics. That is, global environmental degradation is fundamentally a physical problem, so why should it concern politics? This question can be broken up into two parts: causes and solutions. First, can politics or political economy be considered a cause of global environmental degradation? This is a complicated question because ideas about human organization and their implementation in the world affect the global environment. One might make the case that technological development is the true culprit, and there is certainly ample evidence to support this argument. However, reductive answers like "social organization" or "technology" miss important aspects of the problem. Instead, one might look to the consequences of their closely coupled interactions recognizing that corporate behavior surely influences technology just as technology influences corporate behavior. A better understanding of such interactions might point the way towards more lasting environmental solutions. Many have worked in this area including Hardin (1968) who argues that the solution to global environmental degradation requires a fundamental extension not of technology but of *morality,* the first building block of politics.[16] Technical advancements tend to perpetuate antienvironmental behaviors, but moral or political advancements hold the promise for more lasting environmental solutions. This observation places politics in the same issue space as global environmental degradation.

Disentangling the relationships among morality, technology, and the environment remains an open question. Initial efforts such as Hardin (1968) and Ophuls (1977) make clear that some increase in political control is necessary to curb environmentally deleterious behaviors, but these works have been criticized as draconian and simplistic (Ostrom 1990, 8–9). Nevertheless, their point is tangentially acknowledged by North (1990, 235) who points out that interstate conflict in the post–Cold War era is likely to take the form of increasingly ineffective police actions rather than the meeting of organized armies motivated by well-articulated differences. The use of armies for police purposes can only serve as a stopgap measure because situations that require the outside imposition of order indicate fundamental failures of modern political processes (Brzezinski 1993). If such failures continue to grow and accumulate, the sheer number of ungovernable nations will soon overwhelm the resources and will of those countries still able to field forces capable of transferring adequate resources and imposing order.

Ostrom (1990) provides an extension to the issues raised by Hardin (1968). Her theoretically informed case studies focus on the relationships among individual morality, local political organizations, and local natural environments, which she calls "Common Pool Resources" (CPRs). The motivating question behind this study is simple and at the scale of the individual: Why cooperate instead of free ride? Ostrom (1990, 23–24) draws her theoretical motivation from the same source as Keohane, Simon's bounded rationality. The crucial intuition is that centralized decision making institutions, in order to effectively manage CPRs, require large quantities of high quality information as well as the ability to process it. Although a key assumption of microeconomic rationality, such information and processing capabilities are seldom if ever achieved.

This shortcoming has not prevented centralized institutions—modern, bureaucratic states and Multi-National Corporations (MNCs)—from gaining management responsibility over CPRs. Modern states tend to stress equality of rule application rather than rule efficacy, and it is difficult to separate cooperators who invest in understanding how to use CPRs wisely from free riders who simply overuse CPRs. The inability to exclude free riders from CPRs effectively creates a disincentive for investing in the wise use of CPRs (Ostrom 1990, 58–61). Perhaps the best example of environmental free riders are MNCs who are not tied to any particular location, so they have little incentive to ensure the long-term viability of local CPRs. Maximizing short-term profit motivates MNCs; thus CPRs under MNC control tend to be managed with a view toward short-term economic expediency rather than long-term environmental sustainability, and they suffer commensurately (Ostrom 1990, 205–213). Environmental mismanagement is thus rooted in the fact that short-term costs and benefits are easier to calculate and justify than long-term. While states and MNCs, politicians and businessmen receive the benefits of short-term environmental mismanagement, the costs are pushed off to the local people and fauna who depend on CPRs for their livelihood. For them, the rules of politics are not grounded in economic expediency and individual gain but in expectation stability and environmental viability.

Ostrom (1990) clarifies how individual morality and local political institutions should develop to manage CPRs better. Missing in her analysis is a sense of just how pervasive environmental degradation might be. After all, it is the increase in the scale of human activity that is novel and threatening to the global environment. Choucri (1993a) addresses this lacuna through a series of studies aimed at the global aspects of environmental degradation starting from the same basic assumptions. Both Ostrom and Choucri recognize the need for participatory, political solutions aimed at lowering resource use. Where Ostrom contrasts individual cooperators and free riders, Choucri (1993b, 23) contrasts larger-scale degraders and helpers. Both also look to long-term, contextual solutions as opposed to the short-term and simple. However, Choucri's work is more broad-based and theoretical, which plays out in the work's analytical focus. Choucri (1993b, 3–4) identifies three primary challenges facing environmental politics: linkage (understanding the problem), policy (identifying solutions for the problem), and institutional (creating institutions to support the solutions).

At the highest level, this study seeks to identify and articulate important linkages between the social and natural environments. Using the conceptual structure defined by North (1990), this implies linkages between the third and fourth images, the international and global systems as well as the social and natural environments. Throughout the twentieth century, as the industrial revolution has progressed, natural resource stocks have been depleted (Strong 1993, ix). This is akin to saying that as the world economy or social environment has expanded, the natural environment has receded. To say more about this process is a historical and dynamic endeavor, making it both a question of environmental politics (cf. figure 2.3) and for lateral pressure. Whereas lateral pressure has traditionally accounted for the expansive behavior of countries acquiring colonies before World War II, here it is used to account for the expansive behavior of MNCs acquiring worldwide market share after World War II (Choucri 1993c). The juxtaposition of these seemingly disparate issue areas introduces a range of questions and complications that require articulation and clarification. The remainder of this section is spent developing the strategic application of lateral pressure to the study of environmental politics.

Accounting for and explaining the environmental consequences of world economic growth over the past fifty years is a question for political economy; to account for the linkages between the social and natural environments though, it must be an extended, holistic, and systemic version of political economy. The current economics literature is far more concerned with the effects of environmental regulation on economic growth than the effects of economic growth on the natural environment (Choucri 1993b, 2).[17] The challenge of incorporating environmental issues into the political economy represents another aspect of the linkage challenge. Meeting the challenge requires moving away from state centered analyses and incorporating the actions of MNCs because states are no longer the sole potent actors on the international stage

(Choucri 1993b, 14–16). Politically, MNCs have dominated the world economy since World War II. Environmentally, MNCs are the world's major economic actors and technical innovators, and so they are central to any long-term environmental solution (Choucri 1993c). This realization remains unacknowledged as MNC actions are usually treated as environmentally neutral, as if they had no direct impact on the natural environment. An overly state centered analytic focus explains part of this oversight in that MNCs are not state controlled instruments, but they are significant sources of lateral pressure. Specifically, MNCs are a significant source of national expansion and influence, so it is reasonable to question whether or not MNCs are also a significant cause of global environmental degradation.

The aggregated actions of multiple MNCs, each expanding their own international activities, increase the scale of international trade, thereby increasing the crossing of national borders and of lateral pressure (Choucri 1993c, 206). Therefore, international trade clearly falls under the purview of lateral pressure, but trade can be looked at in several different ways. First, trade can be viewed from the perspective of economic theory. International free trade is governed by the law of comparative advantage, which holds that the efficient exchange of goods leads to optimal outcomes for all parties. MNCs, in reducing transaction costs and responding to market imperfections, are important agents of free trade and comparative advantage (Choucri 1993c, 222–224). Second, trade can be looked at from the perspective of systems theory, which, by providing an alternative viewpoint, can be used to critique the more prevalent economic theory. Systemically, trade introduces additional connections and complexities into the international system. That is, because international trade spatially disassociates production, transportation, and consumption activities, it is a source of economic interdependence, complexity, and uncertainty (Choucri 1993b, 23–24). Spatial disassociation introduces a third perspective on trade, geography, which draws from both the economics and systems perspectives. As domestic resources are depleted, domestic costs rise. Higher domestic costs provide the incentive for MNCs to expand their geographic range, possibly into other states, in search of additional resources (Choucri and North 1993b, 70). Note that MNC actions are not directed by the state, but economic incentives prompt MNCs to operate in the interests of the state. Such interests however have international consequences as stronger states may "environmentally invade" weaker states in search of resources. This argument contrasts strongly with the purported benefits of comparative advantage, all the more so given the potential for strong, core states to push their pollution costs off to weaker, peripheral states (Choucri 1993c, 211–212).

Such interactions introduce another perspective on trade—the political. The politics of trade is normally studied through trade policy formation, a process that places those seeking to foster exports in opposition with those seeking to limit imports. What results is the classic debate between free trade and protectionism. Political economists find this debate somewhat puzzling given the benefits of trade and the inefficiencies of trade policy that tends to limit imports rather than promote exports (Alt et al. 1996, 709–712). This debate plays out as free trade and protectionist coalitions compete within government institutions to implement their views, raising questions of collective action and free riding already developed by Ostrom (1990). Addressing such questions requires the consideration of bargaining processes and coalition scale as free riding is much easier in larger groups than smaller ones. Getting at the trade aspect also requires consideration of *factor specificity,* the ease with which economic factors of production or material resources can be transferred between locations. Factor specificity and scale differentiate the three primary models used by political economists to analyze trade: (1) the Heckscher-Ohlin model, which assumes low factor specificity or easily transferable resources, (2) the Ricardo-Viner model, which assumes high factor specificity or hard-to-transfer resources, and (3) the increasing returns to scale (IRS) model, which is used primarily to explain intraindustry trade (Penubarti 1994; Alt et al. 1996, 692–695).

Geography is implicit within the three trade models as factor specificity implies material movement across space and *scale* implies area of industrial operation and influence. When such movements and influences cross national borders, they constitute international interconnections, lateral pressure, and trade. Such trade has increased tremendously in the twentieth century. Aided by technological advancement, the growth of international trade has outpaced even the growth of world population and economic output (Chisholm 1990, 94–95). As international trade has grown, production and consumption have become increasingly disassociated, thus contributing to the complexity, unintended consequences, and environmental costs that have accompanied recent technological changes (Headrick 1990). The trade models currently used by political economists, however, cannot capture the complexity of the international trading system or the environmental costs that might result from its growth. Each of the three trade models cannot, by itself, explain the empirical record of trade—for each successful application, there is a corresponding failure. Consequently, Alt et al. (1996, 712–714) call on political economists to synthesize, blend, and extend the three models so that international trade may be better understood.

The inability of political economic theory to explain history adequately hearkens back to the neorealism debate surrounding Waltz (1979). With respect to war studies, microeconomic theory was unable to explain the history of a complex international system, and so a more systemic theory, lateral pressure, was used to study interstate conflict by looking at the costs of national growth. With respect to political economy, trade theory is unable to explain the empirical record of a complex trading system, and so a more systemic theory, environmental lateral pressure, is used to study environmental degradation by looking at the costs of national economic growth. Specifically, political economy views trade as a beneficial relationship that contributes to the economic prosperity of all participating nations. The geographic perspective, in contrast, sees trade contributing to systemic complexity, unintended consequences, and environmental costs. For political economy, trade is *costless;* for geography, trade is *costly.* Environmental lateral pressure seeks to explore the tension between political economy and geography by looking to the hard-to-measure costs of trade as well as those of national economic growth.

The items of analytical importance are starting to accumulate. Economics, trade, and the dynamics of growth and expansion are all acknowledged as important, as are linkages between the social and natural environments. Lateral pressure has been identified as the theoretical tool for ordering these items. Yet, the question remains, how can we operationalize these intuitions? Growth of the international system's scale has been identified as a major culprit of global environmental degradation. How can this be measured? Since there exists a significant positive relationship between GNP and energy use (Choucri and North 1993b, 72), and a significant relationship between energy use and the greenhouse gas carbon dioxide (CO_2), Choucri (1993b, 14–15) chooses CO_2 emissions as her environmental indicator, her link between the social and natural environments. This indicator is an excellent choice because it accurately measures technical growth, CO_2 contributes to climate change, and CO_2 data are readily available and of excellent quality (Marland et al. 1989).

Increased atmospheric CO_2 however cannot be tied to any specific location—it applies equally everywhere (Choucri 1993b, 25–26). One of the goals of this study is to make explicit the geographical component contained within lateral pressure theory. Consequently, a different indicator is chosen, one that emphasizes geography and land change—*forest change*, which measures the environmental problem, *deforestation*. The two indicators are actually related as deforestation contributes 23 percent of the world's total CO_2 emissions (Choucri 1993b, 21). More than that, however, deforestation indicates globalization through national expansion and trade (Choucri 1993b, 8). The fundamental intuition here is that expansion of the social environment through globalization and trade impacts the natural environment as measured by forest change.[18] Developing this intuition requires further focusing on measures and methodologies. That is done in the following section.

2.5 Methods and Strategy

The theoretical strategy for a more focused and coherent understanding of global environmental degradation rests on the consequences of economic expansion after World War II. The key question becomes this: How to operationalize the strategy so that revealed intuitions can be empirically tested and verified? This section presents the methodological techniques that will be used throughout the remainder of the study. Topics to be addressed in this section include the motivations for using specific techniques as well as the way multiple techniques fit together to form a larger argument. What will not be discussed is the actual application of these techniques to questions of global environmental degradation as this is done in the subsequent chapters. The strategic goal of this study, and of the methods that follow, is the development of an empirically informed, dynamic model that is both grounded in lateral pressure theory and directed at global environmental issues (Choucri and North 1993b, 130–131). Reaching this goal implies linkage of the social and natural environments.

The three methodologies presented in this section—(1) Geographic Information Systems (GIS), (2) diffusion models, and (3) system dynamics—each contribute to the goal of developing a dynamic model.[19] The map-based GIS analysis presented in section 3.1 introduces the data in a visual and intuitive manner. When combined with a time-series analysis that shows how individual indicators vary over time, chapter 3 provides the context for global environmental degradation since World War II. That is, primary indicators are identified, correlations are noted, and questions are formed. Next, statistical models are introduced in chapter 4 to verify and develop the correlations posited in chapter 3. The diffusion model is a geographical technique that accounts for the complex relationships and multiple connections that exist among countries, thus allowing a single statistical test to apply simultaneously to the national, regional, and global levels. Finally, the system dynamics methodology is used in chapter 5 to create a dynamic model that integrates the findings from the previous chapters.[20] The dynamic model constitutes the logical conclusion of this study as it explicitly links the social and natural environments using lateral pressure theory, and it provides the means for the next step of the research program, policy development (Choucri and North 1993a, 501).

The first methodology to be addressed is GIS. Some think of GIS simply as a way to make maps, but because maps present quantitative information in a visual and intuitive manner, GIS is actually an analytical tool. Generally, GIS provides a way both to organize large amounts of data and to show how it is *spatially related*. That is, GIS shows how the distribution of a variable changes across space by presenting the data within a map or *coverage*. Moreover, GIS allows one to ask spatial questions concerning relationships among the various data distributions. This functionality makes GIS a sophisticated software system (ESRI 1995). This study uses a global coverage to present national-level data.

The capacity to map data distributions should pique the interest of realists as mapping power variables raises the question of capacity differentials between neighboring countries. This suggestion, however, places a multivariate tool in the service of a univariate theory. Instead, GIS is used to develop an inherently multivariate theory, lateral pressure. Specifically, the national ecological profiles showing distributions of population, technology, and resources are transferred from a textual, tabular form to a more visual, map form (cf. table 3.2 and figure 3.1). This allows the analysis to move away from the confining concept of power and even from the relative confines of population, technology, and resources. Using the visualization capability of GIS allows the exploration of a number of different variables. Chapter 3 uses GIS maps to introduce the spatial distribution of several such variables. Instead of random distributions, distinct North and South clusters appear within several of the social and natural indicators. These North and South relationships suggest systemic causal linkages. However, the maps present spatial distributions at one moment in time, and multivariate time-series analysis does not readily conform to map-based data presentation techniques. Thus, the GIS maps of section 3.1 are used to introduce variables and elicit intuitions and questions that motivate additional analysis.

The second methodology to be addressed is the diffusion model, a statistical technique with geographic roots. Pooled regression models do not account for temporal or spatial relationships among the variables. From a spatial perspective, this means that each country is considered equally with all others—its neighbors make no difference. This assumption is problematic because experience shows that a country's neighbors do make a difference as events and trends in one country tend to *diffuse* to its neighbors, correlating values through space as well as time.[21] Spatial autocorrelation is directly analogous to temporal autocorrelation, the ability to predict a value at time t by its value at $t-1$, which violates the assumptions of the standard regression model (Kennedy 1992, chap. 9). Diffusion models, in contradistinction to pooled models, account for spatial autocorrelation using contiguity matrices that introduce a "sense of place" to the analysis (Cliff and Ord 1981). That is, the contiguity matrix tells the estimator which countries touch one another and are neighbors, thereby allowing country to country diffusion processes and trends to be captured and explained. The diffusion model, by explicitly recognizing physical proximity and accounting for spatial transmission, extends a formerly pooled, cross-national analysis to encompass the national, regional, and global scales.

The diffusion model concept is used in chapter 4 as lateral pressure explicitly addresses processes of national expansion that have geographical consequences—that is, neighbors of an expanding country tend to be affected more profoundly than geographically isolated non-neighbors. Because diffusion models capture and account for physical proximity, this makes them especially applicable to questions of lateral pressure. Traditionally, lateral pressure studies have used 2SLS for their empirical analysis. 2SLS solves a system of simultaneous equations that allow for the explicit modeling of feedback.[22] A deep understanding of system motivates the use of both 2SLS and the diffusion model, but they each address different aspects of system. Whereas 2SLS accounts for systemic feedback, the diffusion model accounts for the multiplicity of systemic connections. The diffusion model is chosen over 2SLS for three reasons. First, a diffusion model has never been used in a lateral pressure study, so one is implemented here both because it explicitly addresses national expansion and because it is novel. Second, feedback relationships are explicitly addressed by the system dynamics model developed in chapter 5. Third, the system dynamics model will provide a feedback structure that can be used to formulate a 2SLS model at a later time.

The third methodology to be addressed is system dynamics. The system dynamics model developed in chapter 5 constitutes the logical conclusion of this study because it is *dynamic* (Choucri and North 1993b, 130–131). Like diffusion modeling, system dynamics analyzes important aspects of system, but they are fundamentally different aspects. Diffusion models are linear, spatially oriented, data driven, and temporally static; system dynamics models are nonlinear, feedback-oriented, deductively driven, and temporally dynamic. So the two methods, diffusion and system dynamics, each examine very different aspects of system, which makes them complementary. Specifically, the system dynamics model incorporates the results of the diffusion model as well as the theoretical and experimental insights of the previous chapters and synthesizes them into a coherent whole. The exercise of representing environmental lateral pressure within a system dynamics model provides several benefits. First, relationships and assumptions underlying the previous analysis are made explicit, which fosters a more informed debate and a more fruitful exchange of ideas. Second, placing variables and indicators next to each other and explicitly specifying their relationship helps to ensure theoretical consistency and integrity. Third, the resulting dynamic model allows for scenario analysis, for example, varying the model to see how different assumptions or policies play out. Experimenting on a computer allows for tests that are simply too expensive, risky, or rare in the real world. For these reasons, system dynamics has been a staple of lateral pressure since its inception (Choucri 1978; Choucri and North 1975, 281).

System dynamics brings with it the additional benefit of addressing the limitations of human decision making. Keohane (1986b), North (1990), and Ostrom (1990) all cite the bounded rationality work of Herbert Simon as a fundamental question for international relations. The problem is that the assumptions of microeconomic rationality are recognized as unrealistic, but it remains unclear how better to represent and characterize the way individuals (Simon 1985) and organizations (Simon 1983) actually behave. System dynamics operationalizes bounded rationality by (1) factoring decision making, (2) basing decisions on partial and certain information, and (3) using rules of thumb and heuristics (Morecroft 1983). Each of the three operational insights influences the manner individuals interact with their complex surroundings. First, factored decision making describes the way institutions shape and simplify individual decision making by providing explicit institutional goals, information collection, and standardized processing habits. Thus, institutions provide value by reducing the demands placed on human cognition. Second, that decisions are based on partial and certain data says something important about the way people interact with their surroundings. The world simply presents individuals with too much relevant information, and so people tend to winnow down this data by relying on a few sources that can be trusted. The point here is that individual outcomes are not maximized absolutely; instead, maximized outcomes are traded off against minimized information processing requirements. Third, using rules of thumb and heuristics indicates that people also rely on prior knowledge and experience, a reasonable assumption. After all, if decision makers addressed each problem anew, then expertise and experience would not accumulate, which it surely does.

These three issues, while explicitly developed by Simon, remain implicit within system dynamics (Morecroft 1983). The fundamental lesson of this discussion is that people are boundedly rational when interacting with their complex surroundings, so policies and behaviors undertaken with the best of intentions can go awry when they interact in unexpected ways with other systemic factors at other systemic scales. Thus, unintended consequences and bounded rationality are opposite sides of the same complexity coin.[23] This plays out in environmental politics through increases in the world economy's scale, which are intended to benefit the social environment, but increasingly come at the expense of the natural environment, which again "feeds back" to the social. This fundamental insight underlies the remainder of the study.

2.6 Toward Robust Specification

This chapter, in postulating an organizing question and theory, has covered several topics. The initial question developed in chapter 1, "Does trade help or hurt the environment?" was focused by noticing that environmental degradation in South America, Africa, and Southeast Asia is similar and synchronized. It was postulated that there must be some international-scale process that effectively synchronizes geographically disparate environmental degradation, and trade is identified as one social process with both the scale and potency to affect such change. Working at a more theoretical level, trade is characterized as part of the social environment in contradistinction to the natural environment, and figure 2.1 graphically articulates the challenge of linking them together in the form of three questions—what, where, and how. The first question—"What links the social and natural environments?"—is answered by processes that span the social and natural environments. This question is answered using the structure and terminology provided by environmental lateral pressure that is shown in figure 2.4.

Technical improvements help growing human populations manipulate and control ever larger fractions of the natural environment. Manipulating nature includes the extraction of natural resources that serve human populations. Within environmental lateral pressure, the *population* and

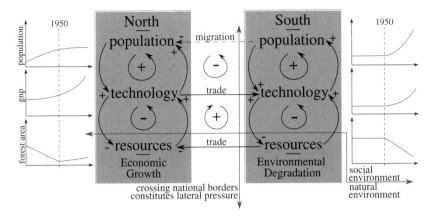

Figure 2.4
Environmental Lateral Pressure. This figure presents the initial, high-level formulation of Environmental Lateral Pressure. Lateral pressure's three master variables—population, technology, and resources—are shown as well as their empirical measures—population, GNP, and forest area. Initial causal connections are offered between the social environment as measured by population and technology and the natural environment as measured by resources. The six graphs to the far left and right demonstrate the dynamics of the system.

technology variables map to the social environment, and *resources* maps to the natural. Figure 2.4, split as it is between the developed North and developing South country groupings, is at the scale of the international and global systems. The global scale is appropriate for two reasons: (1) the increasing scale of human activity degrades global natural resources, or phrased differently, further increases in the international system negatively impact the global system, and (2) environmental politics specifically concerns issues at the interface of the international and global systems (see figure 2.3). As human technology, organization, and infrastructure progresses, ever more land is placed in the service of human wants and needs. International trade aids this process by reallocating resources and increasing the geographical scope of technology. Figure 2.4 shows technology flowing from the developed North to the developing South and resources flowing from South to North. Whereas the relationship between technology and resources is "vertical" (in the "levels of analysis" sense), trade is more "lateral."[24] The extent to which trade processes cross national boundaries is a matter of lateral pressure; the extent to which trade processes impact the natural environment is a matter of environmental lateral pressure.[25]

The second question—"Where do these linkages apply?"—is answered by the concept of geographical location that organizes the international system, the developed North and developing South. However, it remains to be seen just how "North-ness" and "South-ness" are defined. For such concepts to be analytically useful, they must be empirically separable rather than simply "states of mind." Herein, North and South country groupings are separated by their dynamics, as shown to the far left and right of figure 2.4. Note that while population can be measured directly, technology must be measured with the proxy indicator GNP, resources with the proxy indicator *forest area*, and complex international connections with the proxy indicator *trade*. Even though North and South exhibit similar internal structures, the North is marked by economic growth and the South by environmental degradation. As measured by lateral pressure's three "master variables," the North shows a stabilizing population, increasing GNP, and regenerating resources as measured by forest area, and the South shows growing population, growing GNP, and declining forest area. Part of the explanation for this asymmetry is the relative timing of industrialization. The plots show that before 1950, the North had increasing population and GNP and decreasing resources, while the South's three variables were comparatively stable. After 1950, the South began to develop, and its post-1950 profile of increasing population and GNP and decreasing resources looks very much like the pre-1950 North. The GIS maps of chapter 3 test the geographic distribution of the figure 2.4 variables to see if this characterization is empirically verifiable. Additionally, the time-series analysis of chapter 3 tests the characterization of post-1950 North and South dynamics.

The third question—"How should the linkages be expressed and analyzed?"—is answered by the causal connections that link the variables together forming feedback relationships, topics more fully developed in chapters 4 and 5. Chapter 4 begins by testing and verifying some of the causal connections shown in figure 2.4. The polarity of these connections, as denoted by a positive or negative sign next to the arrowhead, corresponds to the parameter polarities derived from the regression tests.

The diffusion model is used to account for the multiple trade connections that exist among nations within the international system. These statistical tests provide the empirical foundation for and validation of the figure 2.4 model. Note that multiple causal connections combine to form the complex, feedback relationships that link a system's structure with its dynamics, and these too have polarity as denoted by the positive or negative sign surrounded by the circular arrow at the center of the loop. However, such relationships can never be subtle enough to perfectly model reality. Models remain abstractions regardless of their sophistication (Sterman 2000, 846–850); so researchers can only hope to capture dominant system linkages and generate insights that compare favorably with other models. With this in mind, the figure 2.4 model is developed and tested more fully in chapter 5. Although the technical details have been postponed until later, it is enough to realize here that understanding complex, feedback systems is central to answering the questions surrounding neorealism, trade theory, and environmental lateral pressure.

Something else has been accomplished through this discussion, something that might escape notice if not explicitly presented and developed. The chapter began with a discussion of Plato's *Republic* and the three-tiered structure of the soul, polis, and cosmos. In progressing from the three images of Waltz (1959) to the four of North (1990), Plato's structure has, in a sense, been rediscovered. The three image structure offered by Waltz (1959)—the individual, the state, and the international system—is wholly social with the soul mapping to the individual and the polis bifurcating into the centralized state and the anarchic international system. With no image corresponding to the cosmos however, Waltz's structure remained unfinished. North's fourth image, the global system, completes the structure insofar as there now exists an image that corresponds to Plato's cosmos.

The increasing salience of the natural environment has consequences beyond abstract questions of international structure and dynamics—it affects the very conceptualization of international politics. Thucydides has long been a favorite of international relations scholars. Beyond the similarities between Athens and the United States, Sparta and the Soviet Union, and the Peloponnesian and Cold Wars, the Melian dialogue succinctly captures something fundamental and inescapable about interstate conflict and the human condition: "... the standard of justice depends on the equality of power to compel and that in fact the strong do what they have the power to do and the weak accept what they have to accept" (Thucydides 1972, 402). This insight speaks to the enduring character of the international system as well as the realists' continuing concern with power, not because it is desirable, moral, or correct, but because power is the universally recognized currency within the anarchic international system. While this truism has long been recognized, it became particularly salient during the Cold War. This period of superpowers, summit meetings, and nuclear weapons stressed centralized states, diplomacy, and constraints imposed on growing nations by the international system. Enemies were easy to identify during the Cold War as were the stakes of conflict, global thermonuclear war.

Many of the international system's features that remained so constant for so long throughout the Cold War have changed suddenly with the collapse of the Soviet Union. No longer do two superpowers vie for control in the far reaches of the globe. Instead, the more prosaic concerns of food, shelter, medicine, and culture stage their comeback. In places where necessities are in short supply, conflict arrives again in new, less-organized forms (Homer-Dixon 1991, 1994; Homer-Dixon, Boutwell, and Rathjens 1993; Kaplan 1993, 1994). Thus, the most apropos passage of Thucydides for the post–Cold War era may no longer be the Melian dialogue, but "The Plague":

Athens owed to the plague the beginnings of a state of unprecedented lawlessness. Seeing how quick and abrupt were the changes of fortune which came to the rich who suddenly died and to those who had previously been penniless but now inherited their wealth, people now began openly to venture on acts of self-indulgence which before they used to keep dark. Thus they resolved to spend their money quickly and to spend it on pleasure since money and life alike seemed equally ephemeral. As for what is called honour, no one showed himself willing to abide by its laws, so doubtful was it whether one would survive to enjoy the name for it. It was generally agreed that what was both honourable and valuable was the pleasure of the moment and everything that might conceivably contribute to that pleasure. No fear of god or law of man had a restraining influence. (Thucydides 1972, 155)

Given such social difficulties, states find it difficult if not impossible to maintain order. Consequently, the impact of diplomats and state power fades as the imperatives of necessity and economy assert themselves.

Realism's power makes way for lateral pressure's population, technology, and resources, and national expansion becomes constrained not by the social environment and international system but by the natural environment and global system. Enemies are hard to characterize in the post–Cold War era as are the stakes of global environmental degradation.

Meeting these daunting challenges requires an essentially different intellectual stance. In some ways this stance is quite new, but in other ways it constitutes a return to a more fundamental and basic set of questions: "I would say that if political scientists or social scientists don't ask themselves how human beings fit into the larger context of existence, if they don't ask themselves how do we somehow or other fit into the large world in which humanity is planted, I think there is something missing" (Cropsey 1990, 43). North's fourth image, the global system, explicitly acknowledges the importance of humanity's place in the world. Moreover, it implies a subtle shift in how politics is conceived. No longer do political issues begin and end with human concerns and the social environment. Environmental politics, by focusing on the nexus of the international and global systems, incorporates the concerns of the natural environment recognizing that these issues are political as well.

3
Contextual Imperatives

Environmental degradation plays an increasingly important role in politics within and among nations. The increased prevelence of environmental discussions reflects a recent and fundamental change in international politics that traces back to the fall of the Berlin Wall in 1989. The fall of the Berlin Wall constituted a sea change for international politics, most notably because it ended the Cold War, the state of mutual hostility that existed between the United States and the Soviet Union from 1950 to 1990. The Cold War did not affect international environmental politics per se, but its end did provide an opportunity for environmental issues to become more visible on the world stage. This release from constraint redefined security in that heretofore neglected topics like the state of the global environment could now be considered within the rubric of security studies (Ullman 1983; Mathews 1989; Walt 1991).

Since the end of the Cold War, the language of international relations no longer centered on the struggle between the communist East and capitalist West for the hearts and minds of the nonaligned, Third World nations. With the collapse of communism and the world's subsequent embrace of market capitalism, attention was directed away from the tension between East and West and towards that between the developed North and the developing South, definitions that presuppose the desirability if not inevitability of development. The tension between North and South is not new and was in no way invented to replace the Cold War. The split had been smoldering in the United Nations' General Assembly since at least the 1972 U.N. Conference on the Human Environment in Stockholm (Tolba and El-Kholy 1992, 742–745), and went back as far as the nonaligned movement of the 1950s. However, prior to 1989 the superpowers remained cloistered away in the Security Council. When the Cold War ended, the diplomatic table cleared making room for issues of interest to the South.

This increased salience of the North and South axis, as well as the emergence of global environmental issues, is exemplified by the 1991 trade dispute between the United States and Mexico (Bhagwati 1993; Williams 1993). In this dispute, a General Agreement on Tariffs and Trade (GATT) dispute-settlement panel ruled for Mexico and free trade and against the United States and the unilateral imposition of environmental standards. The conflict began when the United States placed an embargo on Mexican tuna that had been caught in purse-seine nets that kill dolphins; the use of purse-seine nets was outlawed in the United States. The GATT panel ruled in favor of Mexico arguing that a country's trading rights cannot be unilaterally terminated because of the process used to make a product, in this case tuna.

While there are many notable facets to this dispute, two are of particular interest. The first bears on the relationship between North and South. The border between the United States and Mexico provides a stark and immediate contrast between a developed and developing nation (Graham 1996), so it is natural that such a dispute should turn up here with the United States representing the North by seeking to impose environmental protection on a developing country that cannot afford the same standards. The second regards Mexico's reaction and the implied tradeoffs between environment and development. Southern nations have traditionally resented and rejected environmental instruction from the North, especially when the instruction is backed by the unilateral cessation of trade that generates the capital necessary to repay international debts and fund development.[26]

Tradeoffs between development and environment lie at the root of global environmental degradation. Such issues are complicated, thorny, and increasingly salient on the international stage. For these reasons the United Nations Council on Environment and Development (UNCED) held its 1992 Earth Summit in Rio de Janeiro, Brazil. UNCED—or more popularly, "Rio"—provided a forum to confront the dilemma between economic development and environmental degradation, as the two have been historically coincident (Halpern 1992). National policy in the South is torn between requesting wealth redistribution from the North to fund development and the outright rejection of the North's imposed development paradigm as inappropriate, undesirable, and perhaps even infeasible. Attempting to create, articulate, and implement an improved set of alternatives remains a constant goal and continuing project in the post-Rio era, a project made more difficult by the increasingly interconnected international system (Saurin 1993). Not only is cause and effect ever more difficult to disentangle, but the policy levers available for states to address problems of environment and development are increasingly unclear.

Here we have seen the tension between economics and the environment born out in a more tangible, immediate, and political fashion. Currently there exists a dissatisfaction with the present world economic system, especially in the South, while the North tends to be more dissatisfied with the degrading state of the global environment. These ill-defined anxieties became manifest in Rio as concerned groups and individuals came to grips with these problems in an international forum and tried to formulate policies that would lead to lasting solutions. This remains an ongoing project, and this chapter investigates these issues by reviewing the empirical evidence. In doing so, the traditional purview of international politics—that is, the social environment—is expanded to encompass interactions between the social and natural environments. Specifically, the thesis *that the expansion of the social environment has come at the expense of the natural* is investigated. Addressing the consequences of physical environmental degradation ultimately requires an investigation into their social causes.

While recognizing that correlation is not causation, the data are reviewed to provide context for the subsequent statistical analysis. First, a series of geographic maps are presented to introduce the primary empirical measure used in the analysis—*deforestation,* or more accurately *forest change*—and to give an idea of its spatial distribution. Second, a temporal trend analysis is presented to illustrate how deforestation has progressed and developed over time. This helps us to understand if the trends that are assumed to be coincident truly are. Following this chapter, a statistical analysis is presented as the next step beyond noting that coincidence is determining what supportable causal inferences can be drawn.

3.1 Geographic Analysis

In crafting a study of global environmental degradation, it is easy to be overwhelmed by the details. First, there are so many aspects to the global environment that it is impossible to characterize the system or its behavior in an analytically complete fashion. Second, the causes of global environmental degradation are themselves so varied and localized that it is hard to characterize them generally. While their importance is certainly acknowledged, such details can mask two overarching analytic factors, *scale* and *spatial distribution*. Scale addresses the increasing globalization of economic processes coupled with the concomitant degradation of the natural environment. While environmental degradation has always been present to some extent through history, it is the recent and increased scale of economic activity, especially during the latter part of the twentieth century, that is of major concern (Williams 1990). Economic activity and environmental degradation do not occur everywhere in the same way, and the spatial distribution concept acknowledges this global variegation. While scale and spatial distribution are readily acknowledged as relevant and important to global environmental degradation, their representation and analysis remains an ongoing challenge.

The empirical portion of this study develops an analytically rich methodology for expressing the scale and spatial distribution of environmental degradation, and in so doing we begin to articulate the posited connections between the social and natural environments. Initially, this takes form as a map-based or geographic analysis. GIS is able to present large amounts of spatial information in a graphical and quickly understandable manner (ESRI 1995). The data are presented in an inductive, empirically based, and cross-sectional form that represents only a single moment in time, rather like a methodological "snap-shot." Despite the lack of dynamics, such an analysis quickly conveys the flavor of an issue, a feeling for the problem.

3.1.1 Foundations

This investigation begins with a study of the international system. Choucri and North (1975) and Choucri, North, and Yamakage (1992) address conflict within the international system not by focusing on the traditional measure of international relations, power, but on the more disaggregated and lower-level explanatory variables of population, technology, and resources. These variables can then be used to characterize different nations according to their profile, that is, their relative endowments of these three "master" variables. The concept of profile was extended to environmental analysis by Choucri and North (1993b) and North and Choucri (1996), in which they respectively profile countries according to their 1986 and 1991 global share of population, GNP (a proxy for technology), and area (a proxy for resources; cf. Kindleberger 1962). While the last two variables—GNP and area—do not correspond exactly to what they measure—technology and resources—it is both plausible and probable that they are positively correlated. The world's population, GNP, and area are then summed, and each country's share is generated, thus allowing for comparisons among variables that are not strictly equivalent. This exercise yields the six profile groups in table 3.1.[27]

The profile definitions of table 3.1, when applied to cross-national data for population, GNP, and area, yield the groupings shown in table 3.2. The exercise proves useful in that certain relationships among countries emerge, but it is hard to recognize more general patterns among the data. Recasting information in a more visual format, such as a map, aids visualization as larger scale spatial patterns among countries emerge more easily and naturally. To accomplish this, the groupings are mapped using GIS.

Table 3.1
Ecological Profile Definitions

Group 1: *Area > Population > GNP*
Group 2: *Population > Area > GNP*
Group 3: *Population > GNP > Area*
Group 4: *Area > GNP > Population*
Group 5: *GNP > Area > Population*
Group 6: *GNP > Population > Area*

Source: Choucri and North 1993b, 73.
Note: The six profiles are based on a country's global share of population, GNP (a proxy for technology), and area (a proxy for resources). For Groups 1 through 3, population share is greater than GNP. For Groups 4 through 6, GNP share is greater than population.

Table 3.2
1991 Ecological Profile Groupings

Group 1 (45):	Albania, Argentina, Belize, Bhutan, Bolivia, Botswana, Brazil, Burkina Faso, Cameroon, Central African Republic, Chad, Chile, Colombia, Congo, Equatorial Guinea, Estonia, Gabon, Guinea, Guinea-Bissau, Guyana, Iran, Kazakhstan, Kyrgyz Republic, Laos, Madagascar, Mali, Mauritania, Mozambique, Namibia, Nicaragua, Niger, Panama, Papua New Guinea, Paraguay, Peru, Solomon Islands, South Africa, Suriname, Tanzania, Turkmenistan, Uruguay, Vanuatu, Venezuela, Yemen, Zimbabwe
Group 2 (49):	Benin, Bulgaria, Burundi, Cambodia, Cape Verde, China, Comoros, Costa Rica, Cote d'Ivoire, Dominican Republic, Ecuador, Egypt, Ethiopia, Fiji, Gambia, Georgia, Ghana, Guatemala, Haiti, Honduras, India, Indonesia, Jordan, Kenya, Kiribati, Latvia, Lesotho, Malawi, Malaysia, Mexico, Morocco, Nepal, Nigeria, Pakistan, Romania, Rwanda, Sao Tome and Principe, Senegal, Sierra Leone, Sri Lanka, Swaziland, Syria, Tajikistan, Togo, Tunisia, Turkey, Uganda, Uzbekistan, Western Samoa
Group 3 (20):	Bangladesh, Czech Republic, Dominica, El Salvador, Grenada, Hungary, Jamaica, Lithuania, Mauritius, Moldova, North Korea, Philippines, Poland, St. Kitts and Nevis, St. Lucia, St. Vincent and the Grenadines, Thailand, Tonga, Trinidad and Tobago, Ukraine
Group 4 (5):	Australia, Canada, Iceland, Oman, Saudi Arabia
Group 5 (7):	Bahamas, Finland, New Zealand, Norway, Sweden, United Arab Emirates, United States
Group 6 (28):	Antigua and Barbuda, Austria, Bahrain, Barbados, Belgium, Bermuda, Cyprus, Denmark, France, Germany, Greece, Hong Kong, Ireland, Israel, Italy, Japan, Luxembourg, Malta, Netherlands, Portugal, Puerto Rico, Qatar, Seychelles, Singapore, South Korea, Spain, Switzerland, United Kingdom

Source: North and Choucri 1996, 40–41.
Note: The specific countries in each of the six profiles (154 total countries; values in parentheses denote group subtotals).

The most striking feature of figure 3.1 is the empirically grounded split between the North—North America, Europe, and Japan—and the South—Latin and South America, Africa, and Southeast Asia. While there are six profiles, the relationship between global population and GNP share

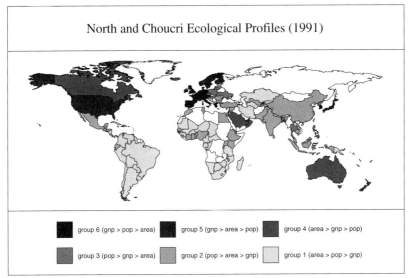

Figure 3.1
North and Choucri Ecological Profiles. This map shows the spatial distribution of the table 3.2 countries. The darker countries are more developed; the lighter countries are less developed. Note that while the same information is conveyed—countries and profile numbers—this format is visually more intuitive (North and Choucri 1996, 40–41).

bears most directly on the divergence of national profiles. Specifically, the three darker profiles—groups 4, 5, and 6—all have a greater share of GNP than population, while the lighter profiles—groups 1, 2, and 3—all have a greater share of population than GNP. Thus, there exist empirical referents located wholly within the social environment that describe "north-ness" and "south-ness" more accurately than their placement on the globe. Exceptions, however, do exist: rich, southern nations include Australia, New Zealand, and the wealthy, oil-producing states of the Middle East; poor, northern nations include the former Communist bloc—Eastern Europe, the Balkans, Cuba, and North Korea. To test the robustness of these profile groupings, the definitions of table 3.1 are applied to the data from World Bank (1995), which generates a separate set of profiles.[28] Figure 3.2 reveals the same broad relationships between North, defined as those countries with a greater global share of GNP than

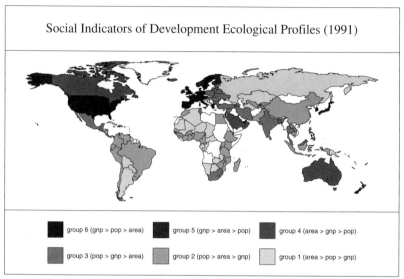

Figure 3.2
Social Indicators of Development Ecological Profiles. This map shows the spatial distribution of the Choucri and North profiles as defined by Social Indicators of Development (World Bank 1995) dataset. It demonstrates the same development patterns as figure 3.1.

population (groups 4, 5, and 6), and South, those with a greater global share of population than GNP (groups 1, 2, and 3). This result gives additional confidence that the profile concept, part of the foundation of lateral pressure, as developed herein reveals something basic about the structure of the international system.

3.1.2 Extensions

Recall that lateral pressure has been extended to the natural environment. So too, the relationships among social variables extend and engender global environmental degradation. The spatial distribution of global environmental degradation is significantly affected by the structure of the social environment. Figure 3.3 shows the geographic distribution of Choucri's (1993a) empirical referent, carbon dioxide per person or CO_2 per capita.

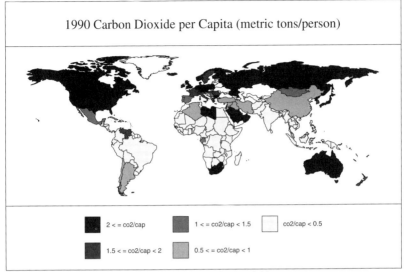

Figure 3.3
1990 Carbon Dioxide (CO_2) per Capita. High CO_2 per capita discharges occur in high GNP per capita countries as petroleum fuels development (Marland et al. 1989).

CO_2 makes a good environmental indicator for several reasons. First, CO_2 comes from the burning of fossil fuels that power development. Note the strong visual correlation between the high GNP countries of figures 3.1 and 3.2 and the high levels of CO_2 per capita in figure 3.3. Thus the northern countries, those that have a higher global share of GNP than population, produce the most CO_2. As technology or GNP grows, so does CO_2 output. Second, CO_2 effluents impact relations among states in that effluents generated in one location become part of the atmosphere where they are dispersed, and their consequences are felt everywhere. States desiring to control such effluents have limited ability to do so because they are incapable of controlling the actions of other states or the atmospheric flows that cross their borders. Consequently, governments cannot protect their populations from the environmentally deleterious actions of others. Third, carbon and oxygen are important components of the natural environment.[29] The burning of carbon-based fuels constitutes both a modification to the natural environment as well as a linkage between the social and natural environments.

CO_2, while acknowledged as a key environmental indicator, is nevertheless problematic. First of all, CO_2 emissions are so omnipresent, invisible, intangible, and apparently benign that people have a difficult time appreciating their potential impact. Second, the scale, scope, and time frame of CO_2 induced global warming is so vast that it is beyond the experience of the nonscientist to appreciate. Third, there exists considerable uncertainty regarding global warming, so the average person is forced to accept the problem's very existence based on faith or trust in technical experts who disagree even among themselves. It should be noted that none of these criticisms denies that global warming is serious. Instead, they make a more political point that CO_2 emissions lack the visual and visceral appeal necessary to motivate action as they lack immediacy or accessibility to the human scale.

3.1.3 Innovations

In this work, in contrast to Choucri (1993a), forest change generally and deforestation specifically are used as the indicators of environmental degradation. This choice is made for several reasons. First, deforestation is itself an important contributor of greenhouse gasses: the burning of fossil fuels in the North generates about 74 percent of the CO_2 released, while deforestation in the South generates another twenty-three percent (Mathews 1992; Choucri 1993b, 22). Clearly, the burning of fossil fuel makes a greater carbon contribution, but deforestation contributes more directly to the loss of species, to a sense of moral outrage, and to the interruption and eradication of biogeochemical cycles. Second, the study of CO_2 and global warming reveals the mismatch in scale between global environmental degradation and politics. Once CO_2 is released it becomes part of the invisible atmosphere, a resource shared equally by all. Not only is an individual unable to see the problem, but that person has difficulty even conceiving its vast scope or the infinitesimal impact of his contribution. Deforestation, in contrast, is much more localized and is thus more visible and understandable. Consequently, deforestation is more at the scale of the individual and of politics. Third, this localization of environmental degradation bears on matters of security redefinition where

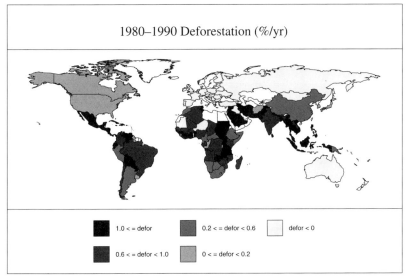

Figure 3.4
1980–1990 Deforestation. This map shows the spatial distribution of deforestation. When compared with the countries in figures 3.1 and 3.2, deforestation occurs primarily in the less developed, lower GNP countries (UNFAO 1993, 1995; World Resources Institute 1996).

it appears that investigating the political processes and economic causes behind localized deforestation will prove easier and more fruitful than those that cause more global-scale and atmospheric CO_2 increases. Figure 3.4 introduces current rates of deforestation. The most striking contrast in comparing figures 3.3 and 3.4 is the almost mirror image between industrial effluents in the North and deforestation in the South. Returning to the national profiles of figures 3.1 and 3.2, deforestation takes place primarily in those countries that have a higher global share of population than GNP. That environmental degradation could be driven by the social split between population and GNP is striking. Specifically, we see evidence that the structure of the social environment has consequences for the natural. Here we glean an additional inkling of the fundamental connection that links the social and natural environments—that the expansion of the social environment has come at the expense of the natural to such an extent that the natural is now imminently threatened.

Finally, in noting rates of deforestation, it is important to know how much forest was there initially. For example, a country like Great Britain may register high afforestation rates, but if the country started off with little forest, then the addition of a few more acres will account for a large percentage but small total increase. Similarly, countries with little forest may show the same deforestation rate as another with much forest, but clearly the greater problem lies with the latter country as that translates to more land in transition. Such arguments however are dependent on the time frame—countries currently considered exemplars of afforestation may have had a history of severe deforestation and are only now allowing the land to recover. Moreover, afforestation may only now be possible because of natural resource imports from other, less developed countries. Figure 3.5 shows the spatial distribution of forests, grasslands,

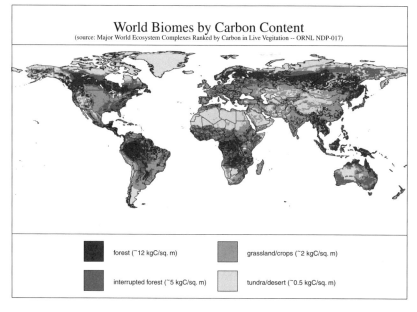

Figure 3.5
World Biomes by Carbon Content. The darker areas denote more forest; lighter areas less. Note that national borders, artifacts of the social environment, show little correspondence to the distribution of forests, artifacts of the natural environment (Olson, Watts, and Allison 1985).

and deserts, each a classic biome or ecological archetype (Rand McNally 1982). Here the complexity and pattern of the global environment is brought into stark relief by the map's smaller scale units. Instead of representing the world with several hundred countries, it is instead represented by 259,200 (360 × 720) half-degree squares. Instead of seeing the world from a political, state-centered perspective, it is seen in terms of plant life and biotic-carbon density. Latitudinal bands of forests and deserts come into stark relief in this projection. The difficult point to appreciate is that there are innumerable relevant variables that could be projected here—for example, rainfall, temperature, fresh water, oil, iron ore, gold, diamonds, molybdenum, etc. Being able to gather, display, and interpret these data remains a methodological challenge for geopolitics and a policy challenge for those concerned with the natural environment. Given the complexity of the global system, to defer policy action until complete information is obtained is impossible—the problem is too big and the data collection task too imposing.

3.1.4 Summary

This section introduced several important concepts through the presentation of five maps. First, figures 3.1 and 3.2 show how national profiles and the comparisons among key national attributes of population, technology, and resources can demonstrate important spatial patterns. Specifically, countries with a greater global share of GNP than population tend to be clustered in the developed North while those with more population than GNP tend to be in the developing South. This relationship between the social variables of GNP and population affects natural measures as well. Figure 3.3 shows that most CO_2 effluents are generated in the North, while figure 3.4 shows that most deforestation occurs in the South. Finally, we see that global-scale data can be displayed in geographical units other than states: natural endowments, like forests, are distributed among states in ways that do not correspond to their borders. These maps however are limited in that they can only present data cross-sectionally, at a moment in time. The following section presents a more longitudinal, time-based analysis that explicitly accounts for change.

3.2 Time-Series Analysis

This section presents a time-series, graph-based analysis in an attempt to establish causal connections between the social and natural environments. As has been established in the previous section, there appears to be geographical relationships between GNP per capita and CO_2 per capita to the North and between population and deforestation to the South. It remains to be seen whether or not these relationships hold over time or whether they are manifest only in that moment captured by the map. Thus, the analysis presented in this section will help insure against specious inferences based on ephemeral correlations. Moreover, this analysis points the way for various statistical tests to be performed in chapter 4. Thus, the goal here is to establish initial connections and correlations among a small set of social and natural variables over a period for which time-series data exist. The exercise explicitly seeks to blend the separate variables presented herein into a larger, more encompassing argument. This analysis does not attempt to identify and explain the history of the multiple variables over an extended time period.[30]

This analysis progresses in two parts, the first focuses on three variables dominated by the developed North and then three by the developing South. Additionally, each variable is addressed at three scales: (1) global, (2) North/South (global split 2 ways), and (3) profile group (global split 6 ways, see appendix B) per table 3.1. Each additional scale gives a better idea of the underlying processes than could any single scale by itself. The North/South and profile scales show that global patterns aggregate many smaller regions, each of which contributes quasi-independently to the process in question. Thus, parts and wholes are contrasted in this analysis, as are structure and process. The global scale gives a general feel for the aggregate process. The more disaggregated measures of North/South and profile group show how variable changes are grouped among similar regions and countries. The scales are limited by the national-level data, although it is recognized that subnational data would reveal valuable spatial patterns were it available. Without regional differences and spatial variation at smaller geographic scales, there would be little incentive for travel and social interaction, which at some basic level constitutes politics. It is the recognition of the aggregation's unstated effects that moves this exercise away from the demographic and towards the political. Finally, these variables are investigated over the time-span covered by the World Bank (1995) report, 1965 to 1992,[31] though it is recognized that the social processes currently causing global environmental degradation have been far longer in the making.

3.2.1 Northern Industrial Expansion

The North historically has been defined as those countries that industrialized first and sought out new lands for colonization, economic imperialism, or free trade. Regardless of the ideological interpretation, it is recognized that the North's historic proclivity for extension and expansion, especially that of Great Britain, Germany, the United States, the Soviet Union, and Japan, has been a consequence of national industrialization and a major factor in the formation of the twentieth-century international system.

However, the definition of "North-ness" is not reserved for those countries with a history of imperialism but for those with *a greater global share of GNP than population*. In a very real sense, GNP or economic output is the driving, dominant variable for this section. The analysis is not limited to economic output however. Three time-series variables are presented in this section—GNP, CO_2, and international trade—and they are presented together as the North dominates all three categories. Moreover, these North-dominated variables are juxtaposed with values for the South, and GNP is additionally discussed in relation to carbon dioxide emissions and trade variables. Visual correlations may be determined among variables and regions that do not follow simply from prior definitions. For this reason, it is worthwhile reviewing these variables to ensure that the data support prior understandings and to explore their temporal and spatial variation.

3.2.2 Gross National Product

GNP is proposed as a proxy for technology in the environmental formulation of lateral pressure theory (Choucri and North 1993b; North and Choucri 1996). The term "technology" here represents the sum of applied knowledge and skills, both mechanical and organizational. In other words, technology consists of the means whereby humans transform and

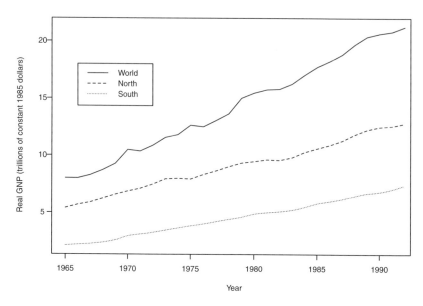

Figure 3.6
Gross National Product (GNP). The GNP data derive from the Penn World Tables purchasing power parity estimates dataset (Summers and Heston 1991). Global GNP grows throughout the period from 1965 to 1992, with Northern countries demonstrating higher GNPs than Southern.

use nature for their own benefit (Headrick 1990, 55), a process that includes but is not limited to nature, people, competition, and economics. Given this expanded perspective, GNP becomes a reasonable substitute for technology. Figure 3.6 shows the sum of GNP for the world, the North and the South. The graph, though it only covers the period from 1965 to 1992, is emblematic of strong worldwide economic growth since World War II despite the small glitch in the early 1980s. Moreover, it is clear that worldwide economic growth is driven primarily by economic growth in the North while the South stagnates by comparison. Northern GNP more than doubled from 1965 to 1992 while southern GNP more than tripled over the same period, but since the South started off from a much lower initial value, its overall economic output continues to lag.

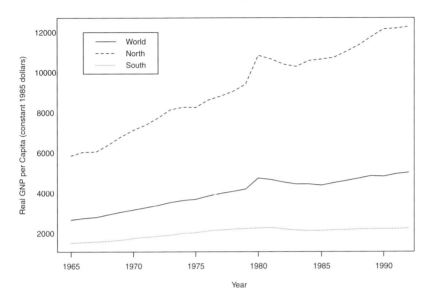

Figure 3.7
Gross National Product (GNP) per Capita. GNP per capita, a measure of individual welfare, grows from 1965 to 1992 for the Northern countries, while Southern GNP per capita remains comparatively static (Summers and Heston 1991).

Figure 3.7 presents mean GNP per capita data for the world, the North, and the South. This data is intended to be interpreted at a more individual level as suggested by Waltz's first image (cf. section 2.2). It becomes immediately obvious that the North has a much higher average GNP per capita than does the South. This can be interpreted by recognizing that GNP per capita is correlated with increased individual resource use and that cumulative resource use results from many individual decisions. Global environmental degradation consequently results from the aggregation of these decisions. Technology increases an individual's access to natural resources, which drives the price of resources down and demand up (Kates, Turner, and Clark 1990, 11). Thus, an increase in technology increases the perceived need for resources as determined by lower prices, while an increase in population increases the absolute need for resources (North and Choucri 1996, 5). When considered from an individual perspective, the North's resource use can be interpreted as a perceived need, while the South's would be an absolute need.[32]

3.2.3 Carbon Dioxide

GNP growth is positively correlated with increases in technology and resource use. Those living within a growing economy however are more likely to remember the new products, experiences, and opportunities that accompany growth. Accompanying economic growth are specialization and surprises, also called *complexity* and *unintended consequences*. Specialization and complexity results from people focusing on doing a particular thing well, which leads to the increasing spatial separation of production and consumption in the world economy (Chisholm 1990). With complexity comes surprises and unintended consequences. These take the form of the externalized costs of industrialization that add up in the natural environment and result in global environmental degradation (Headrick 1990, 58). The example here focuses on the world-wide trade of oil, which has increased markedly along with other commodity shipments since the end of World War II (Kates, Turner, and Clark 1990, 12). The sustained burning of oil, coal, and other fossil fuels has resulted in a major unintended consequence, the buildup of carbon dioxide (CO_2) in the atmosphere.

The flow and use of oil can be analyzed in multiple ways. In physics, oil is a medium entropy resource that gets processed into low entropy fuels that result in high entropy wastes. The burning of these fuels, whether for transportation, heating, or production, powers development and industrialization. What has changed though is the scale of their use. No longer is oil's production and use localized; huge international infrastructures are devoted to its production and distribution, and the effluents from this activity all end up in the atmosphere. Since 1956, the amount of atmospheric CO_2 worldwide has increased from 315 parts per million (ppm) to 350, an 11 percent increase (Houghton and Skole 1990, 401). That an aggregate human activity like industrialization could so alter a shared and elemental resource speaks to the fundamental change in scale that has taken place. No longer does human activity affect nature only in isolated, disassociated locations. The social environment has expanded to the point where aggregate local causes now threaten to alter radically

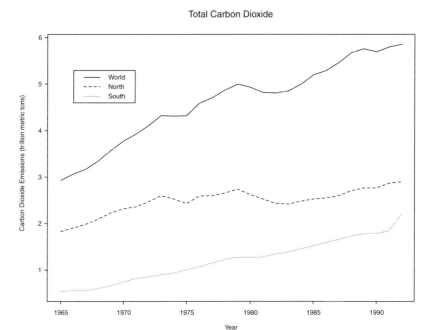

Figure 3.8
Total Carbon Dioxide (CO_2) Output. Carbon dioxide (CO_2) emissions, a byproduct of industrialization and development, increase throughout the period of analysis. Initially the North's contribution is far greater than that of the South, but this gap narrows toward 1990 (Marland et al. 1989).

the global-scale natural environment. Or to phrase it using the parlance of international relations, Waltz's (1959) third image, the international system, has expanded to the point that it now threaten's North's (1990) fourth image, the global system.

Figure 3.8 shows the historical effluent flows that have resulted in the buildup of CO_2 in the atmosphere (a map of CO_2 effluent distribution can be found in figure 3.3). From 1965 to 1992, world CO_2 output doubled from three trillion metric tons per year to six.[33] It used to be generally accepted that an increase in GNP required a corresponding increase in the use of oil or energy, but that has become less true as improved energy efficiency allows for increased economic output using the same or even smaller amounts of energy. Nevertheless, figure 3.8 shows that CO_2

84   Chapter 3

effluents are comparable between North and South and continue to increase on each of the scales shown. Thus, there exists a similar dynamic between North and South in that both regions demonstrate similar development patterns, the only difference being the magnitude of, rather than commitment to development.

Figure 3.9 shows mean CO_2 per capita for world, North, and South. To interpret the figure, recall that increased technology fosters increased consumption of resources by (1) enabling greater access to natural resources, which (2) lowers the price of resources. This holds true for oil and other fossil fuels. The northern countries that have the higher GNP per capita and thus greater technology, show much higher CO_2 per capita

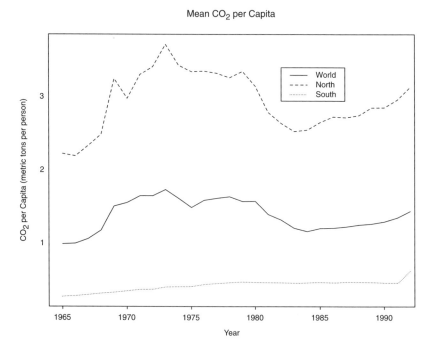

Figure 3.9
Mean Carbon Dioxide (CO_2) per Capita. Northern CO_2 per capita demonstrates a decrease starting in the late 1970s, which is attributable to gains in energy efficiency spurred by the OPEC embargo, reduced supply, and higher energy prices. The South, in contrast, emits consistently less CO_2 per person (Marland et al. 1989).

values than the lower GNP per capita southern countries. This occurs because technology expands the world petroleum stock, the burning of which is the primary source of CO_2 effluents. In some respects, GNP per capita tells the same story as CO_2 per capita. However, where GNP per capita continues to increase, especially in the North, CO_2 per capita has leveled off, primarily after the OPEC price shock of the late 1970s and early 1980s. This leveling off indicates that economic output can increase using the same amount of fuel due to more efficient technologies. In this respect, the relationship between GNP and CO_2 is a bit more complicated than it might appear initially as an increase in GNP does not necessarily imply an increase in CO_2. Finally, although CO_2 per capita is level in the South, its total CO_2 continues to increase due to increases in absolute population rather than relative welfare.

3.2.4 Trade

International trade, the mechanism of physical transfer among nations, is another complicated, multifaceted process that has expanded greatly throughout the industrial era. The tonnage of seaborne freight has increased far faster than world population: in 1900, the world had 1.5 billion people and moved 182 million metric tons of freight; in 1984, 4.7 billion people and 3415 million metric tons of freight moved (Chisholm 1990, 94–95). This rapid increase was made possible through advances in transportation technology, especially the internal combustion engine which not only facilitates but necessitates the trade in oil (Kates, Turner, and Clark 1990, 12). What is traded has recently moved from agricultural goods and commodities to manufactured goods (Chisholm 1990, 98–99), and the growth of world trade has far outpaced world economic output (cf. figures 3.6 and 3.10).

However, not all aspects of trade can be analyzed quite so straightforwardly. The trade process results in disassociations between production and consumption that leads to increased systemic complexity and unintended consequences (Headrick 1990, 58). When products are consumed far from where they are manufactured, it is hard to account for all their lifecycle costs. Moreover, spatial disassociation makes it hard to establish correctly cause and effect with respect to environmental degradation as few government officials or policy analysts can gather enough information and devote enough effort to make such determinations. With so much material moving among so many countries, and with such a complex economic system the result, it is little wonder that the contagions of finance and unintended consequences of global environmental degradation are not well understood. For example, the transportation technologies and systems on which trade depends use large amounts of land. Whether for roads, airports, or railroads, only agricultural activities use more (Headrick 1990, 60). The systemic complexity of trade extends beyond issues of environmental degradation. North and Choucri (1996, 30) state that trade is a modern form of territorial expansion not altogether different from war, and McC. Adams (1990, ix) argues similarly that political power spreads through technology rather than colonialism as the richer countries continue to accrue more power, both physically and politically.

Figure 3.10 reveals that the North leads world imports and exports, just as it led world economic output, with the South lagging far behind. Two other features merit special mention: first, trade decreases just after 1980, presumably because of the OPEC price shock; second, exports are not shown as the import and export curves are quite similar. This is to be expected as one country's export is another's import.

Comparing the total imports and exports graphs is not particularly revealing as the two graphs are quite similar—imports tend to equal exports. Figure 3.11 graphs the trade ratio, *(exports − imports)/(exports + imports)*, to help determine the terms of trade. More exports than imports yields a positive number between 0 and 1, while more imports than exports generates a negative value. The results conform with the previous analysis: trade ratios increase for the northern profile groups indicating a general increase in exports, while the trade ratios for the South fall off indicating a general increase in imports. One exception to this rule is the United States, a developed country that runs large trade deficits, currently on the order of one billion dollars per day (Norris 2001).

3.2.5 Southern Environmental Degradation

If the North has been the traditional generator of geographical extension, then the South has been the area into which the North has extended. Such extension has primarily been driven by the need to supplement diminishing domestic resources. Consequently, natural resources have been

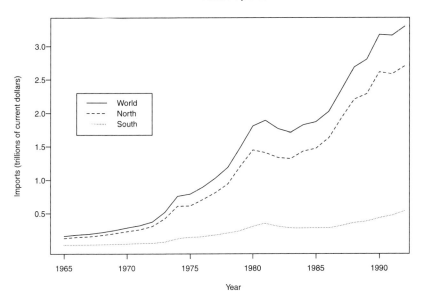

Figure 3.10
Total Imports. Imports only are shown for two reasons: (1) the subsequent analysis uses them exclusively, and (2) the corresponding exports graph exhibits an almost identical response. Note that the value of trade for Northern countries increases dramatically throughout the period of analysis with only a small drop after 1980. Note also that current as opposed to constant values are reported as inflation is difficult to determine and correct for in an international economic context (IMF 1994).

retrieved from the South and brought to the North where they have been turned into manufactured goods both for domestic consumption and export. Historically, the South's importation of northern products resulted in the dramatic decline of southern manufacturing and self-sufficiency (Headrick 1990, 57–58).

However, the definition of "South-ness" is not reserved for those countries that have received Northern imperialism but for those with *a greater global share of population than GNP*. For this section, population is the driving, dominant variable. The analysis is not limited to population however. Three time-series variables are presented—population, forest change, and agricultural land expansion—and they are grouped together as the

Figure 3.11
Trade Ratio. The trade ratio is defined as (exports − imports)/(exports + imports); thus it indicates the relative difference between imports and exports. Positive values indicate that a country exports more than it imports; negative values indicate that a country imports more than it exports. Most countries in the North and South import more than they export, although the terms of trade improve slightly for the Northern countries and worsen considerably for the Southern (IMF 1994).

South dominates all three. However, these South-dominated variables are juxtaposed with values for the North, and population is additionally discussed in relation to deforestation and the expansion of agricultural lands.

3.2.6 Population

Population is one of the three primary lateral pressure variables—the other two being technology and resources. Population, however, is arguably the most important of the three with respect to global environmental degradation. Historically, population has been the driving force behind environmental degradation regardless of the technology employed (Kates, Turner, and Clark 1990, 11). The relationship between technology, here measured through GNP, and population is a complicated and tightly coupled one. First, recent population increases have been driven by technical developments, especially as improvements in medicine, hygiene, and nutrition spread throughout the world after World War II (Whitmore et al. 1990; Demeny 1990). Second, although increases in GNP tend to increase population, population increases tend to undercut advances made in absolute GNP. Figure 3.12 reveals the considerable population growth of the South, and when that growth is combined with the comparatively modest GNP gains demonstrated in figure 3.6, then the result is the flat GNP per capita performance of figure 3.7. Third, viewed from the perspective of the individual, increases in GNP contribute to perceived need, while increases in population contribute to absolute need (North and

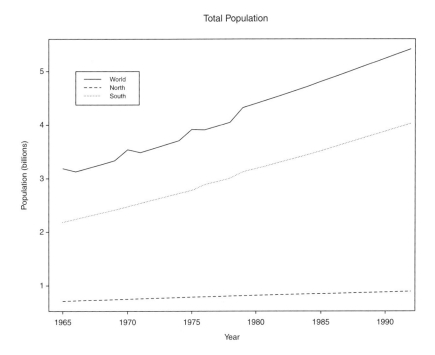

Figure 3.12
Total World Population. World population increases throughout the period from 1965 to 1992, with most of the increase occurring in the South (World Bank 1995).

Choucri 1996, 5). In other words, GNP increases amplify material flows that could be forgone, while population increases amplify basic needs that must be met for living standards to be maintained.

Looking to figure 3.12, we see the continuing trend of world population increase, led primarily by population growth in the South. This trend, here shown from 1965 to 1992, is indicative of a more long-term growth in the global human population that is unique in both scale and speed (Demeny 1990, 47). Until the industrial era, population growth had largely been a regional affair. Now the trend is essentially global as the forces enabling population growth, essentially the technological advances of the industrial revolution, transcend national boundaries.

Figure 3.13 gives the current mean growth rates for the world, South, and North. These growth rates do not necessarily equate to fertility rates

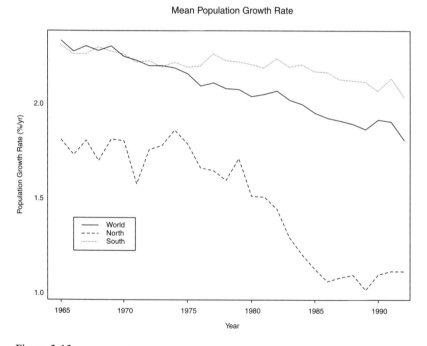

Figure 3.13
Total Population Growth. Population growth rates decrease slightly from 1965 to 1992 with Southern population growth ending at about 2 percent and Northern a little more than 1 percent (World Bank 1995).

as the statistic also captures migration, but international migration is usually small compared to domestic growth (Demeny 1990, 45). The key questions of population growth concern fertility and mortality, or birth and death rates. Conventional demographic wisdom holds that as a nation develops, mortality rates decline as medical care, hygiene, and nutrition all improve. After a time, birth rates fall and a stable population level is achieved (Demeny 1990, 41). The new question, as death rates have fallen in the South, is when will birth rates fall as they have in the North? (Demeny 1990, 51) Though multiple plausible scenarios have been offered, this remains an open question.

Looking more closely at figure 3.13 reveals a southern population growth rate of over 2 percent. This might seem small, but such a growth rate doubles the South's population in fewer than thirty-five years. The North, in contrast, doubles in fewer than seventy years with a growth rate of a little more than 1 percent. However, it should be noted that these growth rates capture migration as well as birth rates, and the North now is a recipient of immigrants. Europe was a source of emigration for centuries. Since World War II though, Europe has become a destination for the world's dispossessed (Demeny 1990, 47). The flow of immigrants from South to North tends to increase northern population growth rates beyond what they would have been due to domestic birth rates alone. Recent wars and other international events have also led to large international population changes. Rwanda experienced a population decrease of 56 percent in 1989 and an increase of 62 percent in 1990 due to its civil insurrection. Kuwait's population fell 39 percent in 1991 due to the Gulf War, as did Qatar's whose population increased 14 percent in 1990. Finally, Israel's population grew by 6.6 percent in 1991 due to the flood of Russian Jews from the former Soviet Union.[34]

3.2.7 Forest Change

Forest change is introduced here as another measure of natural resources and environmental health. As opposed to CO_2, which measures changes in the Earth's atmospheric resources, forest change measures deforestation and afforestation. Forest change is, in fact, consciously introduced in contradistinction to carbon dioxide, though they are related as southern deforestation generates approximately 23 percent of the CO_2 released to the atmosphere (Choucri 1993b, 22). Recognizing that land is ultimately the source of all economic value (Bennett and Dahlberg 1990, 76), this study is concerned more with land change itself rather than the contribution of deforestation to atmospheric CO_2.

Deforestation is used in this study because it is an excellent general indicator of global environmental degradation: "It is possible that the forests illustrate more clearly than anything else how the flows of the Earth, be they migration, trade, or atmospheric circulation, interact with the face of the Earth" (Williams 1990, 197). As with so many other environmentally damaging processes, deforestation had previously been limited to circumscribed regions. It was not until the twentieth century that the process became globalized. Increased technology has affected timber extraction by (1) stimulating demand whether through newsprint, houses, or railroad ties, and (2) increasing the amount of timber it was technically feasible to extract (Headrick 1990, 61). Additionally, land change and deforestation have been fostered by the frontier mentality in which land clearing was considered an almost religious activity. The problem is that frontiers no longer exist—all accessible lands have been occupied (Richards 1990, 177).

Deforestation is an important yet problematic environmental indicator as deforestation data tend to be of questionable quality (Williams 1990, 190–1). The lack of accurate data comes about not due to a lack of commitment or competence but is instead due to the multidimensional and complex nature of forests themselves (Maser 1988). As of yet, forest area cannot be measured adequately or uniformly, although ambitious cross national studies have been attempted (UNFAO 1993, 1995). Figure 3.14 presents the cross-national, time-series data included in World Bank (1995). The data's problematic nature is reflected in the time scale, from 1976 to 1990 as opposed to 1965 to 1992 for the rest of the data set. The most striking aspect of figure 3.14 are the divergent trajectories between North and South: the North is *afforesting*, the South *deforesting*. However, this result is somewhat misleading as deforestation has already been accomplished in North, and much of the afforestation takes the form of tree plantations that may appear like natural forests on satellite photos but are in fact far less sustainable than natural forests (Maser 1988; UNFAO 1995).

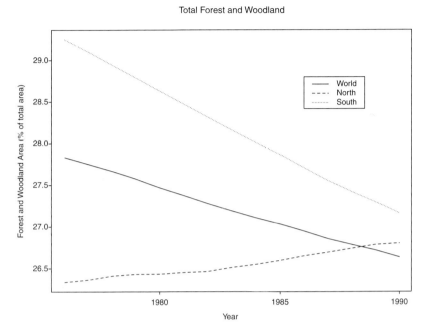

Figure 3.14
World Forest and Woodland Area. The world is losing its forests, but northern forest area is increasing while southern forest area is decreasing. Forest area data tend to be problematic because they encompass large-scale natural processes as opposed to comparatively constrained economic processes. This is demonstrated by the figure's time-scale, which covers from 1976 to 1990 as opposed to 1965 to 1992 for the rest of the World Bank (1995) data.

Turning to the South, although deforestation takes place concurrently with large scale population growth, the historical record indicates that human population increases do not necessarily entail environmental degradation (Whitmore et al. 1990). That is, the relationship between population growth and environmental degradation is complicated, especially in an era marked by globalization. Therefore, it must seriously be considered that the root causes of deforestation may be more global than local. Nevertheless, figure 3.14 provides an empirical reason to view environmental degradation in the North differently from the South because the deforestation difference is one of sign, while that of CO_2 is one of scale. With respect to deforestation, something fundamentally different is going on in the North and South; with respect to CO_2, something fundamentally identical transpires.

3.2.8 Agricultural Land

In looking at and attempting to determine reasons for deforestation, one must answer the question, why? Why does deforestation occur? What motivates such an activity? The need to obtain timber through commercial logging is not, by itself, explanation enough. Land is also cleared for agriculture, the most land intensive of all human activities (Headrick 1990, 61). In many locations, agriculture has changed fundamentally in the twentieth century, as have most of the variables presented in this section. The advent of the petroleum powered internal combustion engine at the beginning of the century closed the distance between farm and city. Machinery and chemicals on the farm also increased the intensity of agriculture. No longer did fields lie fallow; chemical fertilizers were used to make the topsoil work more efficiently with less rest. However, this increased intensification and complexification was not without its unintended consequences. Pesticides like DDT were and are efficient pest eradicators. Unfortunately, their correlative effects tended to disrupt natural ecosystems long after the crops had been taken in from the fields.

Figure 3.15 shows the increase of agricultural lands in the South and the decrease in the North. Thus, the expansion of agricultural land is the inverse of deforestation (cf. figure 3.14). How should one interpret this result? The domestic explanation holds that as populations increase, agricultural productivity must also be increased, or additional agricultural lands must be added in ever more marginal regions to keep up with increased demand for food (Headrick 1990, 62). The international explanation holds that the South's experience entering the world economy has not gone as smoothly as the North's experience creating it. Instead of enjoying the benefits of industrialization, the South has become an agricultural region that exports its crops to the North, an activity that has destabilized and displaced the South's indigenous peoples (Bennett and Dahlberg 1990, 78–79). The North's decreasing acreage of agricultural land would seem to support this argument, but one might respond that the North could just be using advanced agricultural technology to increase yields on ever smaller amounts of land. The real answer is surely some combination of

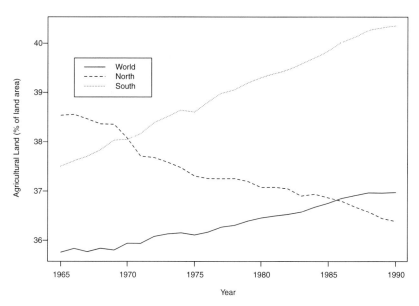

Figure 3.15
World Agricultural Land. Agricultural land is increasing in the South and decreasing in the North. This makes sense as the South expands its agricultural base to feed its increasing population and the North pushes off its food production to Southern countries through trade (World Bank 1995).

domestic and international forces. It is clear though that an inverse relationship exists between forest change and agricultural land and between the developed North and the developing South.

3.3 Conclusion

This chapter has shown that global environmental degradation has helped establish the context of modern international relations by presenting both a geographical and time-series analysis; the former addressed issues of scale and spatial distribution, the latter developed three variables dominated by the North—GNP, CO_2, and trade—and three dominated by the South—population, forest change, and agricultural land. Not only have the variables been examined by themselves, but connections among them have been explored. In doing so, it is difficult to envision and appreciate the other relevant variables that could have been presented: for example, urbanization, percentage of GNP from agriculture and industry, trade mix between agriculture and manufacturing, and fresh water usage. One need only look to Schneider and Boston (1991), Turner et al. (1990), or the eighty-five World Bank (1995) variables to view alternative variable decompositions of the environmental problématique. Thus, a more substantive and far-reaching methodological point is being made, that there exist a virtually unlimited number of intervening, dependent variables (North and Choucri 1996, 4) and that this speaks to the inherent complexity of environmental issues.

This complexity yields several methodological challenges. First, it is important to synthesize and build up systemic relationships among variables in addition to reducing and identifying additional, individual variables. Second, within the multidimensionality of complexity hides the unintended consequences that industrial age technologies have unleashed (Headrick 1990, 64–65). Critiques of capitalism, like the Marxist tradition (Cropsey 1987b), have traditionally taken place within the context of the social environment. Increasingly though, the excesses of capitalism are manifest in the natural. Whether the scale is local or global, harm to the biosphere results conceptually from to the accumulation of unintended consequences. For example, industrial wastes, that grow ever more plentiful, diverse, and toxic are transported ever further away from their source to be more widely dispersed. Oil, the fuel of the industrial era has led, unexpectedly, to the buildup of CO_2 in the atmosphere. And cleared land, so necessary for roads and agriculture, results in deforestation.

In this era of globalization, trade, communication, and effluent transport, unprecedented and ever evolving relationships among previously disparate regions have evolved. This presents significant challenges to current modeling methodologies and the policy prescriptions on which they are based. Just because there is a significant correlation between two local variables, population growth and deforestation for example, this does not mean that an important international-level process is not also at work. Identifying domestic and international causal factors and determining their relative potency remains a research opportunity.

Stated simply, correlation does not equal causation, and a more sophisticated phrasing would stress the importance of multicausality—that one consequence can be brought about by the sum of many causes. With respect to lateral pressure and the environment, the relationships among population, technology, and natural resources are subtle and abstract as the international system is tied together through multiple and multitiered causal connections. Issues regarding global environmental degradation need not remain in a hopeless tangle however. Useful conclusions can be drawn and policy prescriptions formulated—perhaps not at the high level of certainty that has traditionally been demanded though. Understanding policy cause and effect in the international context has become increasingly difficult as the modern relationships among countries confound traditional notions of intention, action, and outcome through the mediation and spatial disassociation of causality (Gereffi 1983; Chisholm 1990; Saurin 1993, 48).

To address such systemic complexities, progress must be made in synthesizing parts and wholes, process and structure. These are hard problems, but environmental politics requires the study of detailed, *low-level* social and natural processes as well as how they aggregate up to affect the larger-scale, global system. Clearly, environmental politics requires a deep understanding of system and a deep understanding of what happens at different levels of analytical acuity. Such questions continue to be addressed and developed in chapter 4.

4
Untangling Complex Linkages: Statistical Analysis

Two broad, geographic patterns of association between the social and natural environments were outlined in chapter 3: that between GNP and CO_2 in the developed North, and population and deforestation in the developing South. However, it remains to be seen whether or not these spatial and temporal patterns are related in a statistically significant fashion. This chapter explores the empirically based, statistical relationships among the previously developed variables within the context of increasing globalization and a more interconnected international system.

This analysis progresses in four stages. First the question, "Does trade help or hurt the environment?," which was introduced in chapter 1, is revisited in the form of a debate between an economist and an environmentalist and then distilled into three prevailing issues: correlation, costs, and complexity. Second, the statistical analysis begins with a single variable, or *univariate*, analysis that explores the spatial relationships concealed within the data. Third, the statistical analysis progresses with a two variable, or *bivariate*, analysis that investigates statistical relationships between the major explanatory variables. Fourth, a many variable, or *multivariate*, analysis is performed. In this section, an explanatory variable is developed that explicitly accounts for trade relationships among nations, Trade Connected GNP (TC × GNP). Testing reveals that TC × GNP explains deforestation better than more widely used explanatory variables, population growth, and GNP per capita.

4.1 Prevailing Contentions

The analytical focus for this chapter is informed by one of the premiere debates to emerge from international political economy at the Cold War's end: free trade versus the environment. Whether the forum is the GATT, the WTO, Rio, or the North American Free Trade Agreement (NAFTA), a fundamental conflict between economists and environmentalists has emerged, persisted, and intensified. The trade and environment issue is introduced through a debate between economist Jagdish Bhagwati and environmentalist Herman Daly.

4.1.1 Key Arguments

Bhagwati (1993) makes the consistent and forceful case that free trade and economic growth are not necessarily incompatible. Economists deeply understand the concept of tradeoffs,[35] and they acknowledge the considerable gray area that exists between conserving the natural environment and ensuring gains from trade. Environmentalists are less comfortable trading nature, says Bhagwati, so they tend to be more emotional, absolutist, and doctrinaire. Economists see nature as humanity's handmaiden, while environmentalists see nature as having value in and of itself. A divergence of views is to be expected when considering that economics is a mature and influential discipline, while the field of environmentalism is comparatively new and developing. However, the differences extend beyond the history of the respective fields and the demeanor of their practitioners. Economists seek free trade and the benefits of specialization guided by the theory of comparative advantage. Their trust is placed in markets, not governments. Environmentalists are less sure of their overall goals and underlying theories, but, according to Bhagwati, increased governmental regulation is certainly part of their agenda.

The fundamental difference between economists and environmentalists is grounded in their divergent views of trade's consequences. Environmentalists are generally suspicious of economic growth and hold that trade furthers such growth, which in turn causes environmental degradation. Bhagwati contends this view is incorrect and argues that trade causes the growth that creates the revenues and taxes that can then be spent by governments to further environmental protection. Society requires funds to spend on nature, and trade among nations helps to generate those funds. How such funds are spent is a political question. Rich nations tend to spend more for cleaner environments than poor nations, so an increase in GNP per capita and trade tends to further environmental protection.

In addition, trade can spread environmentally beneficial technology and knowledge. However, Bhagwati goes on to defend trade for other, non-environmentally based reasons as well.

First, Bhagwati worries that legislation enacted to protect domestic environments will, in fact, be enacted for protectionist, arbitrary, or specious reasons. That is, politicians who create trade barriers ostensibly to protect the environment may actually do so to protect inefficient local businesses. Moreover, even if the legislation is well intended, it could still be enacted without the benefit of science or worse, under the burden of bad science. Thus, it is better to let trade remain free unless an impartial trade dispute panel decides otherwise.

Second, there is the fear that such legislation would be written not to enforce and assure high domestic environmental standards but to impose one's environmental standards on others. For example, the United States' banning of imported tuna because it was caught in non-dolphin-friendly nets constituted an unfair attempt by the United States to impose its environmental standards on Mexico. Such attempts are problematic, argues Bhagwati, because trade theory and modern economics generally is grounded in freedom of action, and restraint of trade hinders freedom.

Third, environmental regulation is problematic because it places the welfare of nature ahead of humanity. Bhagwati expresses concern that the environmental legislation of a rich country like the United States could impact the wages of those living in comparatively poor countries, like Mexico. What's wrong, he asks, in putting people first?

Daly (1993) provides the counterpoint to Bhagwati (1993) by articulating the emerging position of environmentalists. Economists generally, and Bhagwati particularly, contend that when people are rich, the funds will exist to clean the environment. Daly responds that environmental costs are increasing faster than they can be addressed, making society poorer in the aggregate, not richer. This is a subtle and important point that merits further exploration. Economists are emblematic of technical optimism, the view that history is progressive, that any problem is fixable given enough ingenuity and economic incentive, and that it is possible for society to consume large quantities of raw materials without creating long-term environmental and ethical dilemmas (Dobb 1996). These views presuppose reversibility, the notion that what is done can be undone.

Environmentalists appreciate that such is not the case. Some environmental problems are so physically large or technically complex that no human solution is possible besides simply adapting to the situation. This same phenomenon creeps into economics with regard to structural adjustment; if a country's industry specializes according to the exigencies of free trade and the world economy, then does it remain free not to trade? And if it cannot reverse this commitment, then does this connote a fundamental diminution of state sovereignty, a restraint of national freedom? The environmentalist critique contends that economists are so enamored of free trade's conceptual beauty and logical simplicity that they are unable to incorporate such diverse consequences and costs into their thoughts, models, and writings. According to Daly, it is the economists who are blinded, doctrinaire, and increasingly dangerous (Daly 1993, 50).

Daly (1993) interprets free trade far differently than Bhagwati (1993). In fact, Daly disagrees with the very term "free trade," arguing instead that "deregulated international commerce" more closely brings to mind what is at risk given America's expensive economic experiment deregulating the savings and loan industry (Wilmsen 1991). Arguments over terms and underlying theories do not end here. Daly also disagrees with the use of "comparative advantage" as a justification or explanation of free trade. He argues "absolute advantage" better describes the forces at work in the modern world economy. The difference lies in that comparative advantage assumes the immobility of labor and capital beyond national borders. This is no longer true for labor and especially for capital, which in today's global economy is more than fluid, it's hyper-fluid. Billions of dollars can be transferred almost instantly from New York to London and Tokyo, as well a host of other international cities in search of increased returns on investment. In so doing, other goals may be shunted aside such as environmental and labor concerns. The de facto goal of free trade, argues Daly (1993), is the maximization of profit and production without regard to environmental and social costs. He argues that the default economic position should be one not of free trade but of domestic goods for domestic markets as the efficiency gains from free trade are not enough to offset the costs to community and nature. Placing production in closer geographical proximity to consumption helps to identify and account for its costs—costs that continue to mount but do so beyond the limited ability of economists to analyze and appreciate.

4.1.2 Distilling the Issue

As interesting as the debate between economists and environmentalists has been, so far it has produced more heat than light. That is, distinctions clearly exist between the two positions, but the question remains, how can they be reconciled? Instead of concentrating on the institutional or philosophical aspects of this intellectual divergence, this analysis instead focuses on its methodological aspects under the headings of (1) correlation, (2) costs, and (3) complexity. Correlation refers to coincident processes that, when noticed, give rise to questions, positions, and debates. Costs refer to the economists' penchant for adjudicating such debates by turning to the empirical record to determine which positions are supported by facts. Complexity highlights the theoretical limits of counting costs.

The first heading regards the nature of *correlation*. The positions of both economists and environmentalists rest on strong, empirically grounded results. Bhagwati (1993, 42) argues correctly that public policy must be based on logic and facts. He adds that environmentalists all too often rely on vague emotional intimations rather than developed theories and data. His own position is grounded in the results of Grossman and Krueger (1993) among others who find sulfur dioxide (SO_2) effluents decrease as local GNP per capita increases. Bhagwati (1993, 43) concludes from this finding that economic growth is good for the environment: In short, environmentalists are in error when they fear that trade, through growth, will necessarily increase pollution. Environmentalists cannot help but notice the historically concurrent development of international trade, the world economy, and global environmental degradation and feel that there must be some causal linkage (cf. figures 2.1 and 2.4). How to make this argument in a convincing manner remains a challenge.

The second heading regards *costs*. The methodological difficulties for environmentalists become manifest in counting the costs of international trade. Bhagwati (1993) addresses the topic under the heading "social dumping." Environmentalists argue that maintaining high domestic standards, both social and environmental, is hindered when low-priced imports are received from countries with lower social and environmental standards. For example, if high worker benefits and expensive pollution abatement technologies are mandated by domestic law, then relatively inexpensive imports from other countries without such protections unfairly penalize domestic businesses. Consequently, local business and environmentalists feel that a duty should be added to such imports so social standards will not be undercut. Bhagwati opposes such duties for three reasons. First, a diversity of standards among nations is completely natural. Different societies may choose different social standards for a variety of reasons, and import duties should not be applied just because another society makes such a choice. Second, the application of import duties in response to social dumping constitutes an implicit attempt to enforce one set of social standards on another country. Usually this involves a rich country dictating social policy to a poor one, which is undesirable because it restricts the freedom of the poorer country. Third, counting such costs is hard and is open to much interpretation and misinterpretation. Impartial economic decisions can consequently become sullied by politics, self-interest, and protectionism. Bhagwati (1993, 44–46) maintains that such costs simply should not be addressed for the good of the global trading system.

Daly (1993), in contrast, argues that economists send governments inconsistent messages: governments are first told to internalize their domestic costs, and then they are simultaneously told to trade freely with nations that do not do so. This leads to a highly interconnected and globalized trading system in which it is hard to count costs but easy to externalize them. Consequently, the costs of the international system outpace the returns, making the aggregate human population poorer. Moreover, the very scale of the global economic system is no longer sustainable with respect to the natural environment due to the economy's increasing size and mounting costs. This can lead to populations overshooting their carrying capacity, especially if economic costs are geographically separated from their benefits. This inability to accurately count costs proves all the more problematic when one considers that the very basis of economic efficiency requires an accurate accounting of both costs and benefits. If economic benefits are systematically inflated while social and environmental costs are discounted, then there is little reason to doubt Daly's portrayal of international trade as an enterprise driven solely by profit and production maximization.[36]

The third heading regards *complexity*. It has been argued that globalization makes it hard for national regimes to enforce high social and environmental standards. It also makes it hard for economists and environmentalists to count social and environmental costs, but it appears as though environmentalists have more of an interest in doing so. Economists tend to formulate their arguments in terms of a reduced variable set through indifference curves that balance choice between two baskets of goods. "Optimal" results can easily be determined in such a microdimensional mathematical space. Environmentalists, considering more variables over longer time-frames, are faced with a more mathematically demanding set of questions that tend to result in less convincing findings. Additionally, the very systems in question, the social and natural environments, are so varied, multidimensional, and diffuse that it is hard to describe and define such systems let alone account for the costs and benefits resulting from the interaction of their components. Rigorous and consistent definitions for nature, nation, or community do not exist, and the workings of even small-scale ecological or political systems are intricate enough to thwart the most sophisticated attempts at modeling cause and effect. The very belief that environmental damage can be undone with a suitable application of money, labor, and technology is highly questionable. Yet, increasingly sophisticated analytical methodologies, especially those that are computer-based, now allow researchers to address questions that were, until recently, out of reach. This chapter goes some way toward making inroads into this methodological thicket.

4.1.3 Expanding the Inquiry

Light is shed on the aforementioned debate between economists and environmentalists by developing a formal model and empirically testing it. The analysis takes as its starting point the previous findings and correlations that inform each side of the debate. The analysis seeks to locate and identify the international-level costs that have previously escaped notice due to the limitations of the methodologies employed. Specifically, the model developed herein accounts for the systemic interconnections that connect states within the international system. This methodological analysis proceeds in three sections. First, the primary analytic variables are revisited in a univariate analysis. This section supports the notion that the international system is increasingly interconnected and globalized. Second, cross-national connections among the primary variables are explored in a traditional, national-level, bivariate analysis. This section primarily finds that, in a bivariate analysis, there exists a statistically significant relationship between increased GNP per capita and decreased deforestation. This finding directly supports the results of Grossman and Krueger (1993) who found that increased GNP per capita correspond with decreased outputs of SO_2. It also supports Bhagwati (1993), who argues that trade helps the environment because trade contributes to GNP, and increased GNP means fewer trees will be cut. Third, more systemic relationships among nations are explored in the multivariate analysis. This section primarily finds that, in a multivariate, systemic-level analysis, there exists a statistically significant relationship between increased trade-connected, foreign GNP and increased domestic deforestation. This finding explicitly detracts from the results of Grossman and Krueger (1993) and the conclusions of Bhagwati (1993) while bolstering the position of Daly (1993).

4.2 The Simplest View: Univariate Analysis

That the world is increasingly interconnected is commonly voiced. It elicits nods of knowing approval, but the empirical foundation on which this opinion is based remains shrouded in mystery. This section makes explicit the basis for this study's belief in globalization and justifies this chapter's systemic analysis. It does so by focusing on the four primary variables developed in chapter 3: GNP and CO_2 per capita to the developed North, and population growth and deforestation to the developing South.

The analysis begins by providing additional justification for associating GNP and CO_2 per capita with the North and population growth and deforestation with the South. This is done by evaluating the four primary variables as a function of absolute latitude, the simple distance from the equator measured in degrees. Table 4.1 provides the results from these tests. Looking to the signs of the t-values (the ratio of the parameter estimate $\hat{\beta}$ divided by the standard deviation σ), the results tell a simple story: As one moves north or south from the equator, population growth and deforestation decrease, while GNP and CO_2 per capita increase. Because of the relative absence of high GNP and CO_2 per capita countries in the southern latitudes, and the fact that exceptions like Australia, New Zealand, and South Africa tend to be at the relatively high southern latitudes, high GNP and CO_2 per capita countries tend to be thought of

Table 4.1
Absolute Latitude as a Function of the Four Major Variables

Dependent variable	$\hat{\beta}_{|lat|}$	σ	t	pr(> |t|)
GNP per capita	169	20.4	8.29	0.0000
CO_2 per capita	0.0477	0.0062	7.76	0.0000
Population growth	−0.0434	0.0052	−8.31	0.0000
Deforestation	−0.0285	0.0091	−3.13	0.0023

pr(> |t|) denotes the probability that $\hat{\beta}_{|lat|}$ equals the null hypothesis.
Note: These results provide empirical support for the terms "North" and "South." As one moves away from the equator ($0°_{lat}$) and toward the poles ($90°_{lat}$), GNP and CO_2 per capita increase (as denoted by the positive parameters), and as one moves toward the equator, population growth and deforestation increase (as denoted by the negative parameters). Because most high GNP countries are north of the equator, the use of the term "North" for countries having a greater global share of GNP than population and "South" for countries having a greater global share of population than GNP is justified.

as the North, while the high population growth and deforestation rate countries near the equator are referred to as the South (see appendix C for a continuation of this discussion).

Geographical groupings can occur at scales besides the global. Regional variation also exists within the data, which can be investigated with a *distance based spatial weights matrix*. If n countries are being studied, then the weights matrix will be of size $n \times n$. Contained within the weights matrix are 1s and 0s, with 1 denoting another country is within distance d and 0 that it is further away than distance d. If d is 0, then the countries must be physically touching and a *contiguity matrix* results. Such matrices are used in the G_i statistic (Getis and Ord 1992), which provides a measure of spatial concentration:

$$G_i = \frac{\sum_{j=1}^{n} w_{ij}(d) x_j}{\sum_{j=1}^{n} x_j} \forall i \neq j. \qquad (4.1)$$

The variables within equation 4.1 are as follows: n denotes the number of countries or regions studied, G_i denotes an n element vector of results, x_j denotes an n element vector of input values from which the results are derived, and $w_{ij}(d)$ denotes an $n \times n$ element distance or contiguity matrix of 0's and 1's. The resulting statistic, which varies between 0 and 1, allows evaluation of the extent to which large, non-negative values of x are clustered around a country and its neighbors—the larger the value of G_i, the greater the concentration. For cross-national analyses, the statistic provides the practical ability to detect smaller than global-scale spatial clusters of large values.

While G_i introduces notions of space into statistics, there are problems with the measure. First, the results provide no way to evaluate relative clustering. For example, does a value of 0.07 denote a lot or a little clustering? There's no way to tell without comparing the other $n - 1$ results. Second, clusters of low or medium values of x are difficult to measure with G_i. These shortcomings are corrected in the modified G_i^* statistic (Ord and Getis 1995, 289).[37] This more modern statistic allows the use of data with nonzero origins, and variances are developed so that relative clustering can be evaluated. This additional analytic power comes at the expense of mathematical complexity that detracts from conceptual clarity. Therefore, while G_i^* is used throughout the subsequent analysis, a detailed discussion of the statistic is not presented here.

Figure 4.1 gives the probability of spatial clustering for 1990 GNP per capita. Note that probabilities are accorded based on the similarity of nearby values, not on whether such values are low or high, so this map should be compared with the profile distributions of figures 3.1 and 3.2 to understand which countries have high GNP per capita. Three regions of high correspondence are noted in North America, Europe, and, to a lesser extent, Africa. The first two regions, North America and Europe, are high GNP regions, while Africa is a clustered region of low GNP. South America and the Asian countries for which GNP data exist have a more varied distribution of GNP per capita.

Figure 4.2 gives the probability of spatial clustering for 1990 CO_2 per capita. This map should be contrasted and compared with the map of actual CO_2 per capita distribution in figure 3.3 to understand which regions have high and low CO_2 per capita values. This time, three regions set themselves apart: North America, Europe, and the Middle East. In North America, Canada's relationship to the United States qualifies it as a relatively high producer of CO_2, along with Europe and the Middle East.

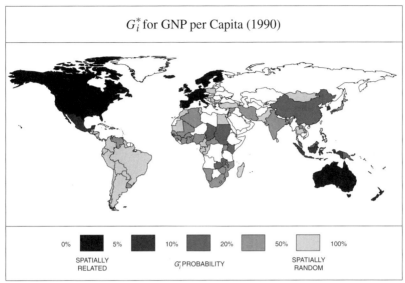

Figure 4.1
G_i^* **for GNP per Capita.** The G_i^* statistic identifies spatially related regions. This figure shows two regions, North America and Western Europe, both of which are notable for high GNP per capita values.

The Middle East is a high CO_2 producer not only because of its high GNP but also due to its production of petroleum products, for example, gas flaring, etc. In contrast, the African cluster centering in Zaire is due to correspondingly low CO_2 outputs.

Figure 4.3 gives the probability of spatial clustering for 1980–1990 population growth. This map should be compared with the profile distributions of figures 3.1 and 3.2 to understand which regions have high and low populations with respect to GNP. Population growth was averaged over ten years to smooth out some of the high 1990 migration outliers. These include negative 4 percent for Bulgaria after the fall of the Berlin Wall, an additional 14 percent for Qatar due to the Gulf War, and an additional 62 percent for Rwanda (following −56% for 1989) due to the cultural conflict between Tutsis and Hutus. Figure 4.3 tells a relatively simple story with Europe demonstrating uniformly low rates of population growth and Africa equally uniform high rates. The Americas and Asia demonstrate more variable population rates.

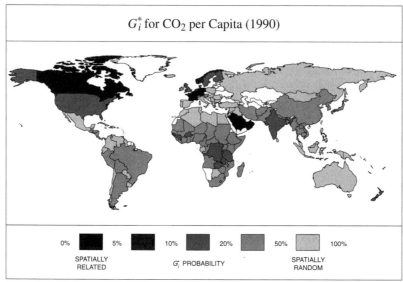

Figure 4.2
G_i^* **for CO_2 per Capita.** CO_2 per capita demonstrates a lack of spatial clustering with similar values found only about the high discharge rates of Canada, France, Germany, and Saudi Arabia.

Figure 4.4 gives the probability of spatial clustering for deforestation between 1980 and 1990. This map should be contrasted and compared with the actual deforestation in figure 3.4 to understand which regions exhibit high and low deforestation values. Three regions demonstrate strong clustering, Central America, Southeast Asia, and Europe. Central America and Southeast Asia do so because of their uniformly high deforestation rates, and Europe because of its uniformly low rates. The two other regions with high deforestation rates, South America and Africa, demonstrate low spatial correlations due to high variation in their deforestation rates.

This section has looked to the global distribution and regional clustering within the four main variables of this analysis: GNP per capita, CO_2 per capita, population growth, and deforestation. Initially, it was established that GNP and CO_2 per capita are geographically clustered to the North and population growth and deforestation to the South. The G_i^*

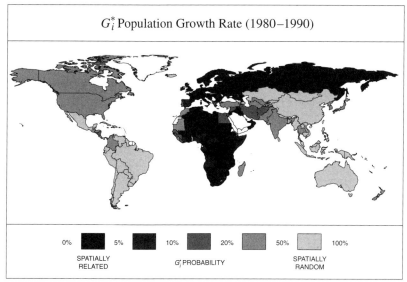

Figure 4.3
G_i^* Population Growth Rate. Two large areas of the world have strongly similar population growth rates: Africa with high rates, and Europe with low.

statistic was then introduced to provide a measure of regional clustering. Such statistics depend upon contiguity matrices that contain information regarding the regional proximity of neighboring countries. The G_i^* statistic was then applied to each of the four primary variables and evidence of regional clustering was found for each. Contiguity matrices will be used again in section 4.4.1 to represent international relationships. Next bivariate correlations within a cross-national, time-series dataset will be explored among the four primary variables examined here.

4.3 The View Less Simple: Bivariate Analysis

Reviewing the change of a single variable over time and space provides useful insights, but so too does analyzing relationships between pairs of variables. In this section, the four primary variables are paired and tested to determine which natural environment variables are influenced by which social environment variables. CO_2 per capita and deforestation are both

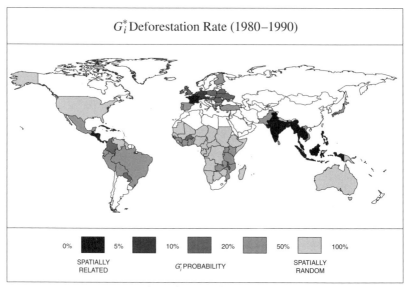

Figure 4.4
G_i^* **Deforestation Rate.** Three regions demonstrate spatial clustering for deforestation: Central America and Southeast Asia, both of which have high deforestation rates, and Western Europe, which has a low deforestation rate.

tested in a pooled fashion as a function of GNP per capita and population growth. Given the previous discussion, it should be expected that CO_2 per capita and GNP per capita would be significantly correlated due to their preponderance in the developed North; deforestation and population growth will likewise be correlated due to their preponderance in the developing South. Although these are fairly simple tests, they give some idea as to whether causal connections exist between the variables. The results of each test are supported by a short table summarizing the regression results.[38]

Table 4.2[39] shows a strong relationship between CO_2 per capita and GNP per capita.[40] This means that, given the empirical record from 1965 to 1992, it is reasonable to expect an increase in GNP per capita will lead to a corresponding increase in CO_2 per capita. This correlation is confirmed by the regression's t-value of almost 80, leaving little doubt regarding the estimates significance as anything less than 5 percent is gen-

Table 4.2
CO_2 per Capita = f(GNP per Capita)

Coefficients	$\hat{\beta}$	σ	t	pr(> |t|)
(Intercept)	−0.191	0.0218	−8.73	0.0000
GNP per capita	300μ	3.76μ	79.9	0.0000

pr(> |t|) denotes probability of null hypothesis; $\mu = 10^{-6}$.
F-stat.: 6387 on 1 and 3698 deg. of freedom, the prob. is 0.

mean(GNP per capita) =	4017	$(1985)/person-year
mean(CO_2 per capita) =	1.39 (1.00)	metric tons/person-year
$\hat{\beta}_1$ × mean(GNP per capita) =	1.21 (0.87)	metric tons/person-year

Values in parentheses denote ratio to mean CO_2 per capita.
Note: CO_2 per capita is strongly and positively correlated with GNP per capita, which is consistent with rich, northern countries that have high CO_2 and GNP per capita.

erally considered good, and this estimate's probability is less than 0.005 percent. Moreover, the magnitude of the estimate is large enough to considerably affect the model's output. Thus, this test confirms what has been suspected, that there exists a close relationship between CO_2 and GNP per capita, especially in the north.

Table 4.3 shows some relationship between CO_2 per capita and population growth. This means that, given the data, there is a slight positive relationship between CO_2 per capita and population growth. However, the magnitude of the estimate is small enough to confirm the lack of a significant relationship. This conclusion is plausible for two reasons. First, so far there has been little evidence to support such a relationship in that CO_2 is thought of more as a northern problem while population growth is a southern problem. Second, CO_2 per capita is calculated by taking a nation's total CO_2 and then dividing through by its population, and so a population increase would tend to decrease CO_2 per capita. However, figures 3.8 and 3.9 show a strong upward trend in total southern CO_2 and a somewhat shallower trend for southern CO_2 per capita. Recall that although the North and South both demonstrate increasing CO_2 outputs, they differ in degree. Change in forest area (cf. figure 3.14), in contrast, demonstrates different slopes for the North (positive) and South

Table 4.3
CO_2 per Capita $= f$(Population Growth)

Coefficients	$\hat{\beta}$	σ	t	pr(> |t|)
(Intercept)	1.18	0.0796	14.8	0.0000
Population growth	0.0623	0.0319	1.953	0.0509

pr(> |t|) denotes probability of null hypothesis.
F-stat.: 3.815 on 1 and 4703 deg. of freedom; the prob. is 0.05085.

mean(Population growth) =	2.04	%/year
mean(CO_2 per capita) =	1.39 (1.00)	metric tons/person-year
$\hat{\beta}_1 \times$ mean(Population growth) =	0.127 (0.09)	metric tons/person-year

Values in parentheses denote ratio to mean CO_2 per capita.
Note: CO_2 per capita is weakly correlated with population growth, in terms of both parameter magnitude and significance.

(negative) indicating fundamentally different forces at play. The table 4.3 test is probably picking up this slight trend of increasing CO_2 per capita and population growth in the South. The barely insignificant probability value of five percent along with a small parameter estimate indicate coincident yet somewhat independent trends.

Table 4.4 shows a strong negative relationship between forest change and population growth. This means that given the empirical record from 1976 to 1991, it is reasonable to expect an increase in population growth will be accompanied by a decrease in forest area. This correlation is confirmed by the regression's t-value of −8.5, which compares well with the general test of significance, as anything less than five percent is considered good and this estimate's probability is less than 0.005 percent. Moreover, the scale of the parameter estimate is large enough to affect the model significantly. This test confirms what was previously postulated, that there exists a close relationship between forest change and population growth, especially to the South. However significant this relationship is, it is recognized that many intervening factors and variables complicate the causal linkages. For instance, additional tests show an increase in agricultural land is also strongly correlated with forest area decreases.[41]

Table 4.4
Forest Change = f(Population Growth)

Coefficients	$\hat{\beta}$	σ	t	pr(> |t|)
(Intercept)	−0.0119	0.0152	−0.788	0.431
Population growth	−0.0528	0.0062	−8.51	0.0000

pr(> |t|) denotes probability of null hypothesis.
F-stat.: 72.48 on 1 and 2380 deg. of freedom; the prob. is 0.

mean(Population growth) =	1.97	%/year
mean(Forest change) =	−0.120 (1.00)	%/year
$\hat{\beta}\times$ mean(Population growth) =	−0.104 (0.87)	%/year

Values in parentheses denote ratio to mean Forest change.
Note: Forest change is strongly and negatively correlated with population growth. Forest change is consistent with poor, southern countries having high rates of population growth and deforestation.

The movement of displaced domestic populations into previously unoccupied or sparsely occupied forests is only one mode of deforestation. Another is the more organized, corporate effort to harvest timber. There also exists a gray area between these two, such as when Southern leaders curry favor by giving away free chain saws to the unemployed with promises to buy the wood they cut. This brings up questions of national sovereignty, responsible leadership, and the legitimacy of behaviors engendered by modern states (Turner 1996), all of which further complicate the analysis.

Table 4.5 demonstrates a positive relationship between forest change and GNP per capita, meaning that the data from 1976 to 1991 indicate that an increase in GNP per capita will lead to a corresponding increase in forest area. This test's t-value of almost ten provides strong evidence that the estimate is statistically valid. The magnitude of the estimate is somewhat smaller than those previously encountered. Yet this test still confirms the findings of Grossman and Krueger (1993) and the arguments of Bhagwati (1993) who state that a strong positive relationship between GNP per capita and the environment exists—in this case, more GNP means more forests. This result appears to refute the environmentalists who argue that because global environmental degradation and the

Table 4.5
Forest Change = f(GNP per Capita)

Coefficients	$\hat{\beta}$	σ	t	pr(> |t|)
(Intercept)	−0.199	0.0102	−19.6	0.0000
GNP per capita	16.3μ	1.66μ	9.80	0.0000

pr(> |t|) denotes probability of null hypothesis; $\mu = 10^{-6}$.
F-stat.: 96.07 on 1 and 1933 deg. of freedom; the prob. is 0.

mean(GNP per capita) =	4534	$(1985)/person-year
mean(Forest change) =	−0.120 (1.00)	%/year
$\hat{\beta}_1 \times$ mean(GNP per capita) =	0.0737 (−0.61)	%/year

Values in parentheses denote ratio to mean Forest change.
Note: Forest change is strongly and positively correlated with GNP per capita. Economists contend that trade helps the environment because it contributes to GNP per capita, which leads to more forests.

development of the world economy have occurred simultaneously, the economic system must somehow be responsible for environmental degradation as an increase in GNP per capita means more forests. It also bolsters the central thesis of Bhagwati (1993), that economic growth is beneficial to the environment. Additionally, since free trade furthers economic growth, and since economic growth increases forest area, then free trade by implication increases forest area. These conclusions are tested directly in the following section.

4.4 The Complex View: Multivariate Analysis

In evaluating the causes of global environmental degradation, the statistical evidence supports relationships between CO_2 and GNP to the North and deforestation and population growth to the South. A strong statistical relationship was also found between increased forest area and increased GNP per capita. Does this mean, as Bhagwati (1993) argues, that increased international trade and economic growth reduce environmental degradation?

In answering this question, one must differentiate between domestic and international effects while recognizing that the latter are not adequately represented in the bivariate analyses performed in section 4.3.

Although a range of countries were represented, the bivariate tests were, in a sense, wholly domestic. Consider the last test, forest area as a function of GNP per capita: the forest area of Country 1 is compared against the GNP per capita of Country 1. Repeated across a range of years and countries, this essential pairing ultimately generates the estimate. However, the univariate analysis (cf. figures 4.1, 4.2, 4.3, and 4.4) confirmed that countries do not exist in isolation within the global context—Country 1 is influenced by Countries 2, 3, and 4—but per the geographic perspective, nearer places are more related. In the modern world, globalization among countries is accomplished through a variety of international connections including communication, travel, and diplomacy, but trade surely ranks high among them. Thus it is desirable that globalization, as represented by connections among countries, be incorporated to improve the model. This is done in the next section.

4.4.1 Developing Trade Connected GNP (TC × GNP)

It is true that countries exist within the international system, but it is also true that standard regression techniques do not represent or account for context. International connections are injected into the analysis using the contiguity matrix concept introduced in section 4.2 to account for trade relationships among nations. Recall that the contiguity matrix, $w_{ij}(d)$, previously used to compute G_i and G_i^*, contained 0's and 1's based on the physical proximity of two countries. If two countries, i and j, share a border, then the matrix element w_{ij} contains a 1; if they are geographically separate, then the matrix element contains a 0. In this fashion, it is possible to introduce context and regional variation into a statistical model. Countries are no longer disconnected and decontextualized within the estimate. Instead, a sense of closeness and farness—of context and realism—is introduced. The problem for this analysis is that in an era of increasing globalization, physical proximity may not be the most salient factor in the international system. That is, relationships based on physical closeness may not be a good indicator of trade relationships as countries that are far apart may trade freely while others that are physically close may be relatively trade free.

Figure 4.5 shows an example trading system consisting of four countries, C1, C2, C3, and C4. Each has its own GNP and conducts trade

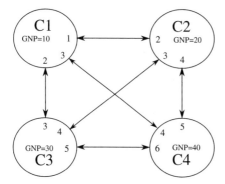

Figure 4.5
Example Trading System. An example trading system of four countries, C1 through C4, is shown. The numbers next to the arrowheads represent imports. Each country has its own GNP and imports goods from others. For example, country C1 imports products with a value of 1 from C2, 2 from C3, and 3 from C4.

with its neighbors, as represented by the value next to the arrow that connects them. This value may represent imports received from or exports heading to the connected country, which does not really matter. What does matter is that the representation remains consistent with whichever choice is made. For the sake of this discussion, trade values will be interpreted as imports. These trade values are then placed into a contiguity-like matrix in equation 4.2:

$$\text{Trade} = \begin{vmatrix} 0 & 1 & 2 & 3 \\ 2 & 0 & 3 & 4 \\ 3 & 4 & 0 & 5 \\ 4 & 5 & 6 & 0 \end{vmatrix}. \tag{4.2}$$

As can be seen, the import values for C1 are placed in matrix row 1, C2 in row 2, and so on. Note that the diagonal, w_{ii}, contains all zeros as a country receiving imports from itself makes no sense.

$$\text{TC} = \begin{vmatrix} 0 & \frac{1}{6} & \frac{2}{6} & \frac{3}{6} \\ \frac{2}{9} & 0 & \frac{3}{9} & \frac{4}{9} \\ \frac{3}{12} & \frac{4}{12} & 0 & \frac{5}{12} \\ \frac{4}{15} & \frac{5}{15} & \frac{6}{15} & 0 \end{vmatrix} \quad \text{GNP} = \begin{vmatrix} 10 \\ 20 \\ 30 \\ 40 \end{vmatrix}. \tag{4.3}$$

The trade continuity or TC matrix in equation 4.3 demonstrates *row normalization*, the dividing of each term in a row by the sum of the row. This is done for several reasons. First, one might want to know not the value of imports received but the percentage of total imports received for a certain country, which is what w_i represents after row normalization. The normalized value thus helps impart a sense of relative size by answering the question, "Is this value a lot or a little?" Second, dividing through by the total imports leaves a dimensionless number that can be compared among years without having to convert to constant currency values, which is hard to do for trade. Third, the dimensionless percentage can be used as a set of *trade connections* and multiplied by other variables to determine international transference effects though the global trading system, in this case GNP. Equation 4.4 shows the trade connection matrix, TC, multiplied by the GNP vector from equation 4.3:

$$TC \times GNP = \begin{vmatrix} 0 & + \frac{1}{6} \times 20 + \frac{2}{6} \times 30 + \frac{3}{6} \times 40 \\ \frac{2}{9} \times 10 + & 0 & + \frac{3}{9} \times 30 + \frac{4}{9} \times 40 \\ \frac{3}{12} \times 10 + \frac{4}{12} \times 20 + & 0 & + \frac{5}{12} \times 40 \\ \frac{4}{15} \times 10 + \frac{5}{15} \times 20 + \frac{6}{15} \times 30 + & 0 \end{vmatrix}$$

$$= \begin{vmatrix} 33.3 \\ 30.0 \\ 25.8 \\ 21.3 \end{vmatrix}. \qquad (4.4)$$

The resulting variable, Trade Connected GNP or TC × GNP, generates a value that depends both on the amount a country trades with its partner and the size of that partner's GNP. GNP per capita is eschewed for this statistic because it matters little whether a trading partner's large GNP is due to high technology and low population or low technology and high population. According to lateral pressure theory developed in chapter 2, each generates considerable incentive for national expansion. Note also that, per figure 4.5, the countries with the lowest GNP have the highest TC × GNP and vice versa. This pattern demonstrates how the economic consequences of high GNP economies might be projected to and accounted for within low GNP countries.

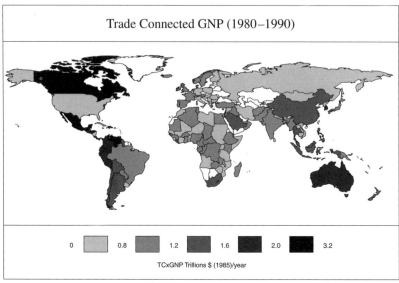

Figure 4.6
Trade Connected GNP. High TC × GNP values denote large amounts of trade with a high GNP partner. Canada and Mexico both have high TC × GNP values due to their trade with the United States, and Venezuela, Columbia, and Peru all have high values due to their trade with the regional power Brazil.

Figure 4.6 shows the spatial pattern of TC × GNP and reveals high values primarily in the lower GNP countries, indicating poor countries that trade with rich ones tend to have the highest TC × GNP. Evaluating TC × GNP with respect to latitude returns a negative estimate indicating that TC × GNP tends to diminish as one moves away from the equator indicating a southern orientation (cf. section 4.2).[42]

Figure 4.7 shows the spatial distribution of the G_i^* statistic for TC × GNP. Note that instead of showing those countries that trade with large GNP nations, the figure shows the center (or nadir) of trading blocs. For this reason, regionally dominant nations like the United States, Brazil, and China are shaded darkly. Other countries at the center of trade-free regions are also shaded darkly, like Yugoslavia. The regional analysis of TC × GNP begins by investigating the measure's regional clustering as spatial correlation violates the assumptions of the standard regression model just as temporal autocorrelation does. Figure 4.7 shows that spatial

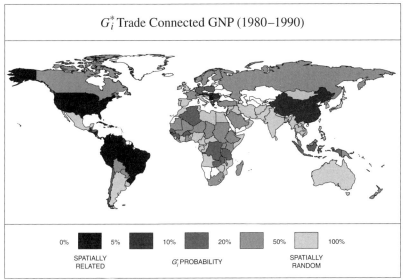

Figure 4.7
G_i^* **Trade Connected GNP.** Countries surrounding the United States and Brazil have similar TC × GNP values, which is a testament to their regional trading prowess. China also emerges as a regional locus of trade.

correlation is present in Latin America, while there is a relative absence of it in Africa, Southeast Asia, and Western Europe.

4.4.2 Explaining Deforestation with TC × GNP

It must be remembered that TC × GNP was not created in isolation. It was created to reflect international trade connections in contradistinction to domestic GNP per capita measures in the context of global environmental degradation generally and deforestation specifically. Thus, it makes sense to test TC × GNP as an explanation for deforestation.

Table 4.6 shows the results of testing forest change as a function of TC × GNP.[43] The interpretation of table 4.6 tells a much different story than that of table 4.5. Instead of the positive relationship between forest change and local GNP per capita, we see a negative relationship between forest change and TC × GNP. The large t-value and relative parameter magnitude indicate that over the time examined, 1976–1991, an increase in either the percentage trade with a high GNP trading partner or an

Table 4.6
Forest Change = $f(\text{TC} \times \text{GNP})$

Coefficients	$\hat{\beta}$	σ	t	pr(> |t|)
(Intercept)	−0.0355	0.0149	−2.38	0.0173
TC × GNP	−0.0979	0.0135	−7.26	0.0000

pr(> |t|) denotes probability of null hypothesis.
F-stat.: 52.72 on 1 and 1933 deg. of freedom; the prob. is 0.

mean(TC × GNP) = 0.935		T$(1985)/year
mean(Forest change) = −0.120 (1.00)		%/year
$\hat{\beta}_1$ × mean(TC × GNP) = −0.0916 (0.76)		%/year

Values in parentheses denote ratio to mean Forest change; $T = 10^{12}$.
Note: Forest change is strongly and negatively correlated with Trade Connected GNP (TC × GNP). Environmentalists contend that trade hurts the environment because it allows rich, northern countries to push their environmental costs off onto their poor, southern trading partners.

increase in a traditional trading partner's GNP results in decreased local forest area for the country in question. This result supports the contention that economic growth in the United States, Western Europe, and Japan contributes to deforestation in Latin America, Africa, and Southeast Asia.

The next step in the analysis is to test the three explanatory variables—TC × GNP, GNP per capita, and population growth—together to see which affects forest change most significantly. This is done in table 4.7. The multivariate model is formulated so that the international (TC × GNP) and domestic (GNP per capita and population growth) explanatory variables are compared against one another. Essentially the same results obtain from the multivariate model: TC × GNP affects forest change negatively, GNP per capita positively, and population growth negatively. All three parameter estimates are statistically significant at the 0.005 percent level. Looking to the estimates themselves, TC × GNP and population growth affect the model most dramatically followed by GNP per capita. Thus TC × GNP outperforms the more traditional explanatory variables, GNP per capita and population growth, in terms of parameter magnitude and statistical significance in explaining global forest change (see appendix D for additional tests of the table 4.7 model including explained variance, temporal autocorrelation, and spatial autocorrelation).

Table 4.7
Forest Change = $f(\text{TC} \times \text{GNP} + \text{GNP per Cap.} + \text{Pop. Growth})$

Coefficients	$\hat{\beta}$	σ	t	pr(> |t|)
(Intercept)	0.0112	0.0228	0.4904	0.6239
TC × GNP	−0.121	0.0131	−9.26	0.0000
GNP per capita	13.9μ	1.78μ	7.81	0.0000
Population growth	−0.0391	0.0063	−6.21	0.0000

pr(> |t|) denotes probability of null hypothesis; $\mu = 10^{-6}$.
F-stat.: 74.14 on 3 and 1925 deg. of freedom; the prob. is 0.

mean(TC × GNP) = 0.935		T$(1985)/year
mean(GNP per capita) = 4534		$(1985)/person-year
mean(Population growth) = 1.97		%/year
mean(Forest change) =	−0.120 (1.00)	%/year
$\hat{\beta}_1$ × mean(TC × GNP) =	−0.114 (0.95)	%/year
$\hat{\beta}_2$ × mean(GNP per capita) =	0.0630 (−0.52)	%/year
$\hat{\beta}_3$ × mean(Population growth) =	−0.0770 (0.64)	%/year

Values in parentheses denote ratio to mean Forest change; $T = 10^{12}$.
Note: The parameter magnitude of *TC × GNP* compares favorably with the more traditional explanatory variables, *GNP per capita* and *Population Growth*, as the negative contribution of *TC × GNP* outweighs the positive contribution of *GNP per capita*.

To recap, this section presents, develops, and tests a new variable, Trade Connected GNP (TC × GNP), which is calculated by cross multiplying trade connections with the GNP of a country's trading partners. Thus, the interconnections within the international system are modeled explicitly as opposed to more standard pooled models where such connections remain implicit. TC × GNP is then introduced into a multivariate model to explain forest change along with GNP per capita and population growth. It compares well. Unlike GNP per capita, TC × GNP contributes negatively to forest change. That is, where GNP per capita increases lead to forestation, TC × GNP increases lead to deforestation. Moreover, TC × GNP generates a regression parameter of larger relative magnitude so that it contributes more to forest change than GNP per capita detracts from it. The contribution of TC × GNP to forest change is actually on the same scale of the more generally acknowledged causal variable, population growth. This result provides explicit contrast between an international-level independent variable, TC × GNP, and two domestic-level variables, GNP per capita and population growth. So the multivariate model explicitly compares international and domestic causal variables within a common, two-level game type format (Putnam 1988). Additional tests performed in appendix D reveal the same pattern—the negative influence of TC × GNP on forest change is more statistically significant, influential, and stable than the positive influence of GNP per capita. That the variable TC × GNP remains statistically significant in the presence of more widely recognized and generally accepted national-level variables, GNP per capita and population growth, means that international-level trade processes contribute significantly and negatively to forest change.

4.5 Conclusion: Revising the Contentions

This chapter began with a review of the economist's and environmentalist's contrasting views on free trade as articulated by Bhagwati (1993) and Daly (1993). Bhagwati maintains that free trade and economic development are good for the environment, while Daly maintains that they contribute to social and environmental degradation. Their positions were distilled and organized under the headings of correlation, costs, and complexity, and the three headings will be used again to organize the conclusion. The time-based analysis of chapter 3 showed that it is prudent to postulate connections among growth, globalization, and global environmental degradation given that they developed over the same period. Thus, the concerns of environmentalists are not without empirical foundation. However, correlation is not causation, and Bhagwati's insistence on logic and facts as opposed to the vague emotions sometimes offered by environmentalists is advice well taken.

This chapter has attempted to resolve these two positions through the application of statistical methods that better count the costs of global economic development. The univariate analysis reintroduced the four main variables of the previous chapter: GNP per capita, CO_2 per capita, deforestation, and population growth. It confirmed that the first two are primarily present in the North with the latter two in the South, and it showed that all four variables demonstrate regional clustering. The measures of environmental degradation—CO_2 per capita and forest change—were then tested against the two social variables—GNP per capita and population growth—in the bivariate analysis. CO_2 per capita was found to vary positively with GNP per capita, and forest change negatively with population growth. Bhagwati's position was confirmed when forest growth was found to vary positively with GNP per capita, indicating more wealth means more trees. In the multivariate section, the trade connected GNP variable, TC × GNP, was developed and tested against forest change along with GNP per capita and population growth. Trade connected GNP was found to be environmentally negative, statistically significant, and more influential than GNP per capita. Stated simply, more trade means less trees. The use of trade connected GNP allows the model to count explicitly the costs of trade, and these costs impact forest change negatively.

How do these results affect the debate between Bhagwati and Daly? Let us first recall that forest change is an important and arguably optimal environmental indicator (Williams 1990, 197). Therefore if global forests are affected negatively by trade, then it is reasonable to infer that the aggregate global environment is also affected negatively by trade. Second, let us recall that economists, in criticizing the antitrade position, assert that environmentalists are in error when they argue that trade, through growth, hurts the natural environment (Bhagwati 1993, 43). Given that GNP and trade in combination have been shown to decrease forest area, and given that forests are a reasonable measure of environmental health, then Bhagwati's statement appears to be false: Environmentalists do have reason to fear trade and growth. Disproving the economists does not mean that the environmentalists are wholly correct. Whether international trade increases competition and lowers prices by increasing efficiency and lowering labor and environmental standards is not addressed by this model. However, the model does indicate that trade may serve as a mechanism for the exportation of environmental costs by high GNP countries. The model also makes clear that trade and growth affect the environment negatively, which is enough to refute the economists' claim.

Simply disproving a claim is not enough though. It must also be asked, what gave credibility to the argument in the first place? The proffered answer is that the complexity of the international system was not properly taken into account in previous analyses, and figure 4.8 helps to explain

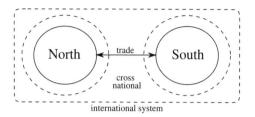

Figure 4.8
Competing Model Perspectives. The statistical results presented herein derive from a more encompassing perspective of relations between North and South within the international system. Cross-national studies, as denoted by the two inner dashed circles surrounding North and South, focus on the countries themselves and examine them equally in aggregate without taking into consideration the relationships among them, as defined by geography, trade, or something else. The outer dashed rectangle represents the wider perspective offered by TC × GNP, which brings trade relationships into the model.

this position. The two circles labeled "North" and "South" represent groups of northern and southern countries. These country groups are connected by a series of trade relationships as indicated by the two-headed arrow labeled "trade." The difficulty arises with respect to the analytical perspective employed. A pooled, cross-national estimate, as indicated by the two dashed circles surrounding North and South and demonstrated by table 4.5, captures only the national-level processes at work in each country, but such models miss international-level connections like trade. This happens because each country is treated alike with no recognition of regional variation or systemic complexity. The introduction of trade connected GNP widens the perspective from the two dashed circles to the outer dashed box, from the cross-national to the international system, allowing for the costs of trade to be explicitly counted. Reliance on models that lack this wider, international perspective explains how previous analyses regarding the relationships among environment, globalization, and trade were mistaken.

5
Exploring Complexity: System Dynamics Analysis

The relationship between economic wealth and environmental health is complicated. Rich nations tend to have good environments, while poor nations evince the worst natural degradation. This has led economists to conclude that the best way to ensure the lasting health of the global environment is for all nations to become wealthy. So goes the argument, since trade contributes to global economic growth, and growth benefits the environment, trade is therefore good for the environment.

This policy prescription—that trade helps the environment—oversimplifies and conflates their true relationship. This study has shown that GNP increases in a country's trading partners lead to increased deforestation for the country in question (see table 4.7). This result supports the opposite conclusion, that the richer countries are, in fact, not more environmentally benign; they instead push their environmental costs off onto poorer nations. Should this proposition prove true, then globalization, trade, and world economic growth are not the means to an environmentally sustainable future; they instead become the means to an unsustainable existence at the environmental expense of the world's poorer countries.

Chapter 5 explores this question by developing a simulation model that evaluates, interprets, and synthesizes three analyses: the time-series analysis of chapter 3, the statistical analysis of chapter 4, and the economic concepts of Ricardo's comparative advantage and Malthusian overshoot and collapse. Section 5.1 recalls the analytical tension between the optimistic theory of comparative advantage and the relatively pessimistic scenario of overshoot and collapse. Section 5.2 examines how new, computer-based representations influence the balance of this tension. And section 5.3

128   Chapter 5

develops the simulation that blends and reconciles the previous analyses. Specifically, it demonstrates just how the North pushes its environmental costs to the South.

5.1   Recalling Analytic Tensions

Analytic tensions between trade and the environment are not new; they extend back to the very beginnings of economic theory. If Adam Smith is the father of neoclassical economics, then his two putative protégés, Thomas Robert Malthus and David Ricardo, are the fathers of the environment and trade debate (Samuelson and Nordhaus 1992). Malthus takes up the pro-environment position by disagreeing with the economists of his day who held that large, growing populations necessarily lead to wealth and prosperity. He argued that populations tend to expand until they are limited by available food, especially as populations tend to increase far faster than food supply (Malthus 1976 [1798], 23). Malthus reasoned that in America, where food and land were plentiful, population doubled every 25 years or about every generation. Thus, population grows geometrically as demonstrated by the series 1, 2, 4, 8, 16, 32, ... Food production, in contrast, tends to grow arithmetically, or according to the progression 1, 2, 3, 4, 5, 6, ... Given these two trends, eventually there will not be enough food to go around. Malthus consequently maintained that growing populations will eventually exhaust all arable land attempting to grow food. Within Malthus's theory, the social and natural environments are not only represented explicitly but interact directly in that an inevitable lack of natural resources will ultimately limit human population growth. Malthus's answer to this dilemma is moral restraint and a cessation to reproduction until resources are available (Malthus 1976 [1798], 130–132). In this fashion, social cognition, volition, and choice are introduced into an otherwise deterministic process.[44]

Ricardo (1911 [1817]), working from Malthus's economic assumptions, reasoned further that as a population increased and the food supply dwindled, wages would be driven down to the subsistence level. His proposed solution was not the extension of morality but the importation of cheap, foreign grain through free trade and the theory of comparative advantage. Whereas some economic theories are justly criticized as ques-

Table 5.1
Comparative Advantage Example

Production capacity	Great Britain	Trading partner	
cloth	$C_{GB} = 4 \frac{bolts}{person}$	$C_{TP} = 1 \frac{bolt}{person}$	
wheat	$W_{GB} = 2 \frac{bushels}{person}$	$W_{TP} = 1 \frac{bushel}{person}$	
Isolation			Totals
	$C_{GB} \times 50 = 200$	$C_{TP} \times 50 = 50$	250_{bolts}
	$W_{GB} \times 50 = 100$	$W_{TP} \times 50 = 50$	$150_{bushels}$
Specialization			
	$C_{GB} \times 60 = 240$	$C_{TP} \times 25 = 25$	265_{bolts}
	$W_{GB} \times 40 = 80$	$W_{TP} \times 75 = 75$	$155_{bushels}$
With Trade			
	207 $\xrightarrow{33}$	58	265_{bolts}
	102 $\xleftarrow{22}$	53	$155_{bushels}$

Note: Trade increases the wealth of all trading partners according to the principle of comparative advantage. In this example, Great Britain creates cloth at the rate of 4 bolts per person and wheat at the rate of 2 bushels per person; its trading partner does so at the rate of 1 bolt per person and 1 bushel per person respectively. Note that Great Britain has an absolute advantage making both cloth and wheat. Without trade and with 50 people working in each industry in each country, Great Britain produces and consumes 200 bolts of cloth and 100 bushels of wheat; its trading partner produces 50 bolts of cloth and 50 bushels of wheat. If each country specializes by moving workers into the industry in which it has a comparative advantage, then Great Britain will produce more cloth and its trading partner more wheat. After Great Britain trades its excess cloth for its partner's excess wheat, both countries have more cloth and wheat than they did without trade.

tionable, obvious, or ineffective, comparative advantage is recognized as true, nonobvious, and effective (Lane 1998b). Comparative advantage allows for greater consumption on the part of all who participate in trade as can be seen in the example presented in table 5.1. Imagine two countries, Great Britain and a Trading Partner, both of whom make two products, cloth and wheat. Each country has one hundred workers who are equally split between cloth and wheat production. With isolated economies, this yields 200 bolts of cloth and 100 bushels of wheat for Great Britain and 50 bolts and 50 bushels for the Trading Partner. Specialization yields different results. If Great Britain moves ten workers from wheat to cloth production and the Trading Partner moves twenty-five workers from cloth to wheat, then Great Britain will make 240 bolts and 80 bushels and the Trading Partner will produce 25 bolts and 75 bushels. With specialization, total cloth production increases from 250 to 265 bolts and wheat increases from 150 to 155 bushels. If Great Britain trades 33 bolts of cloth for 22 bushels of wheat from the Trading Partner, then both countries will have more cloth and wheat to consume. Both populations are materially better off through specialization and trade. Moreover, a country need not have an absolute advantage to benefit. Great Britain is more efficient than the Trading Partner in producing both cloth and wheat, and yet both countries benefit from specialization. It is the comparative advantage that results from the diversity of natural and technical endowments that fosters the increase in consumption and material wealth.

Historically, Ricardo and the theory of comparative advantage seem to have won the day over Malthus and the theory of natural limitation. Great Britain's role as leader of the nineteenth-century world economy placed it in an ideal position to export high technology manufactured goods like cloth and import its foodstuffs and other raw materials. Even so, both Malthus and Ricardo underestimated the twentieth century's technical advances that would seemingly refute Malthus' core assumptions. Over the past hundred years, and especially the last fifty, increases in economic output, world trade, and globalization have far outpaced population increases, which were themselves significant, thereby increasing real wages (Samuelson and Nordhaus 1992, 695). Moreover, Malthus predicted that wealth increases would lead to subsequent increases in population, but population growth has slowed most in precisely those regions that have experienced the greatest economic gains, Europe and North America. Modern economists maintain that resources need not limit population given suitable applications of technology.

It would be premature however to state baldly that Malthus is wrong and Ricardo is right. The debate's longevity underscores that the relationship between environment and trade is a fundamental, core concern of the international economy. New analytic methods and advanced computers have allowed such tensions to be examined from fresh perspectives, a topic that is examined in the following section.

5.2 Representing Analytic Tensions

Although the debate between trade and environment is in some ways well-worn and classic, new technical advances can breath new life into old questions. The notion of natural limitation of social growth reappears in the influential study *The Limits to Growth* (Meadows et al. 1972).[45] Modern analytical techniques update Malthus by enabling analysts to connect more variables thereby creating more complex models. *Limits* focuses on the interactions among five separate global factors—(1) population, (2) agricultural production, (3) natural resources, (4) industrial production, and (5) pollution—and notes five significant trends—(1) accelerating industrialization, (2) rapid population growth, (3) widespread malnutrition, (4) nonrenewable resource depletion, and (5) environmental deterioration. It concludes that given present trends of exponential population growth, industrialization, pollution, food production, and resource depletion, the limits to global population growth will be reached within 100 years, resulting in sudden and uncontrollable declines in births, population, and industrial capacity. Thus the essential lessons of Malthus are revisited, this time at the global scale.

At the analytical heart of *Limits* is *system dynamics,* a computer-based simulation methodology used to represent and evaluate systems more mathematically complex than those studied by Malthus and Ricardo.[46] System dynamics models are generally useful in uniting, coordinating, and arranging theories and data from divergent disciplines. For this specific problem, system dynamics reconciles the competing demands of the social and natural environments in terms of global economic growth and environmental capacity. The *overshoot and collapse* concept is central both to Malthus and *Limits* and will be used here to introduce a range of system dynamics' modeling concepts (HPS 1990, chap. 9). The analysis of a system dynamics model takes place in three steps: (1) structure, (2) feedback, and (3) dynamics. Beginning with *structure,* figure 5.1 shows a system that exhibits a basic overshoot and collapse response. System dynamics models are composed of four basic elements: (1) *stocks,* the

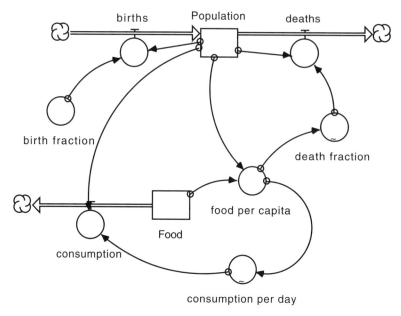

Figure 5.1
Overshoot and Collapse System Structure. This is a simple system dynamics model, also known as a "stock and flow" model. The model's stocks, which represent collections of things, are *Population* and *Food;* the flows, which allow things to flow into and out of the stocks, are *births, deaths,* and *consumption.* The other circles are called "converters," and the arrows "connectors" (see figure F.1 for the model's nonlinear relationships).

rectangles that contain *Population* and *Food* rather as a bathtub contains water; (2) *flows,* the circles connected to the rectangles by pipes that either fill up or draw down the stocks; (3) *converters,* the circles connected only by arrows; and (4) the arrows, which are called *connectors.* Figure 5.1 contains two stocks, Population and Food, which take initial values 2 and 100, respectively. System dynamics models create differential equation-based models, just like those that are used to explore chaos theory (cf. appendix A). The fully specified equations for the overshoot and collapse model are presented in section E.1.

Connected to the two stocks are three flows: *births* and *deaths* are connected to Population and consumption to Food. The births flow is defined as the converter *birth fraction,* defined simply as the constant 0.2, multiplied by Population. So each year, 20 percent of the population gives birth thus increasing Population. The deaths flow decreases Population, and it is defined as the product of Population and *death fraction*. The death fraction converter is more complicated than birth fraction: where birth fraction is a simple constant, death fraction's value is determined by *food per capita*, the result of Food divided by Population. Similarly, the *Food* stock is drawn down by *consumption*, the product of Population and *consumption per day*, which is in turn determined by food per capita. The converters death fraction and consumption per day are nonlinear tables as described generally by figure 1.1 and defined specifically in appendix F (Converters defined by tables are denoted by a tilde in the circle).

The structure of Figure 5.1 forms a web of causal, *feedback* connections as depicted by the *causal-loop diagram* of figure 5.2.[47] The causal connections are denoted by arrows that "feed back" on themselves forming circular loops. Both causal connections and feedback loops have polarity. A "plus sign" next to a causal arrow denotes change in the same direction. For example, the causal arrow from *food/capita* to *consumption* in figure 5.2 is positive because figure F.1 shows that consumption per day increases as food per capita increases. They change in the same direction—more food implies more consumption. Conversely, a *"minus sign"* denotes change in the opposite direction. The causal arrow from food/capita to deaths in figure 5.2 is negative because figure F.1 shows

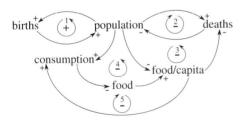

Figure 5.2
Overshoot and Collapse Feedbacks. Individual causal relationships aggregate up to form larger-scale, system-level feedbacks. As described in figure 1.2, there are two types of feedback, self-reinforcing, positive feedback, and goal-seeking, negative feedback. This figure shows the system structure of figure 5.1 as a set of five feedbacks: one positive and four negative. These feedbacks drive the simulation's dynamic response.

that death fraction decreases as food per capita increases. They change in opposite directions—less food means more deaths.

Feedback relationships, as denoted by the circular arrows in the center of the loops, also have polarity. Feedback loop #1 in figure 5.2 shows that more births lead to more population, and more population leads to more births. Such a relationship is a reinforcing or positive feedback and is denoted by a plus sign in the center of the loop. Unrestricted positive feedback can lead to exponential growth, so called because not only does the process continue to grow, but the rate at which it increases also grows. Positive feedbacks however tend not to grow without bound. Population increases lead to increases in both births and deaths, hence the positive signs on the causal connections. As more births increase the population, more deaths decrease the population. Thus, the arrow from deaths to population has a negative sign indicating that the two values vary in opposite directions. Feedback loop #2, relating population and deaths, is a balancing negative feedback as the process seeks to maintain a steady value, whereas the positive loop between births and population seeks to grow without bound. Thus negative feedbacks check the growth of positive feedback. Positive and negative feedback loops can be determined by counting the negative causal connections around the feedback loop: An odd number makes the feedback negative; zero or a positive number makes it positive. There are three other negative feedbacks in figure 5.2: loop #3 connecting population, food/capita, and deaths; loop #4 connecting population, consumption, food, food/capita, and deaths, and loop #5 connecting food, food/capita, and consumption. Each negative loop seeks to "balance" the system, in part by working to limit the positive feedback between births and population.

The limiting character of negative feedback can be seen in the dynamics of the system shown in figure 5.3. Population grows rapidly, but as it does, Food decreases. When Food finally runs out, Population falls. The dynamics are generated not by historical data, but by the internal logic of the model. The model shows that if a population grows too large for its food supply, then the death rate will increase, possibly to the point of population collapse. Phrased more technically, positive feedback may dominate for a time, but it will eventually be limited by a negative feedback response. This is essentially the lesson of Malthus and *The Limits*

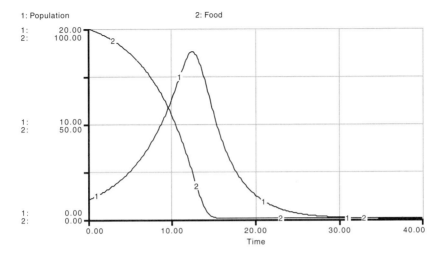

Figure 5.3
Overshoot and Collapse Dynamics. The nonlinear and feedback causal relationships described and depicted in figures 5.1 and 5.2 combine to yield these dynamics. The two stocks from figure 5.1 are graphed: *Population* and *Food*. Population increases so long as there is Food. When the food supply is exhausted, that is Food is 0, Population collapses until it too is 0. The model's initial growth is the "overshoot," and the subsequent Population drop the "collapse."

to *Growth*. However, such lessons are controversial and have prompted many critiques and countercritiques. *Limits* was criticized as too aggregative, as glossing over important regional differences, and as averaging away the concerns of the poorest countries (Cole 1977). Critics argued that the world is too complex to capture in one simple model. Perhaps the most damaging criticism has been that the predictions made by *Limits* did not occur. Anticipated shortages of minerals and oil not only didn't materialize, but new discoveries increased supplies and drove down prices (Economist 1997a).

Since *Limits* was published, global modelers have learned several lessons. First, predicting the long-term timing of a system as complicated as commodities within the world economy can prove problematic if not impossible. Identifying particular commodities as limiting economic factors is also difficult, especially when substitutes are readily available. But the real problem may be with the predictions themselves. Modelers have discovered that predictions turn out not to be as useful as the learning that takes place in researching and constructing the model (Richardson 1983). The true value of computer simulation takes place in creating a coherent analytical framework, making causal and logical connections explicit, and running multiple scenarios of the resulting system. Such scenarios do not need to predict the system perfectly in order to be useful; they just need to impart more understanding than can be provided by comparable cognitive or spare mathematical models (Sterman 2000, 846–850). Finally, a model should be disaggregated and focused on a specific problem. If the scope of the model is too large, diffuse, or abstract, then its applicability to the real world can become unclear or unrealistic, bringing the model's results into question. Applying a model to a single, specific question has proven an effective way to ensure analytical focus. Given these modeling "rules of thumb" or heuristics, the primary focusing question, "Does trade help or hurt the environment?" is revisited and revised in the next section.

5.3 The Environmental Lateral Pressure Model

The remainder of this chapter develops a model that addresses three specific issues raised earlier in this study. First, the time-series responses of chapter 3 are explained and interpreted by developing a system structure that yields comparable dynamics. Second, this effort provides further support for the trade and environment statistical results presented in chapter 4 by identifying the logic by which the developed North pushes off its environmental costs onto the developing South. Essentially, the model provides a link between the systemic assumptions and dynamic responses necessary to explain the preceding time-series and statistical analyses. Third, the model reconciles the two informing theories—Malthus's overshoot and collapse and Ricardo's comparative advantage. It does so by comparing, balancing, and blending the concepts using Environmental Lateral Pressure (ELP) theory (cf. chapter 2) and the system dynamics simulation methodology. The resulting system dynamics model speaks to the social environment through Ricardo and the natural environment through Malthus using the analytic structure provided by environmental lateral pressure. In addressing questions such as these, the ELP model incorporates lessons learned from previous modeling efforts. First, model results are not presented as predictions but as the conclusion of a larger effort of empirically based research and synthesis. Before the dynamic results are examined, the rationale behind the model's structure and an analysis of its feedbacks are presented. Second, the model is disaggregated into separate regions. Whereas *Limits* was criticized for being too aggregative and for averaging away the concerns of the poorest countries, this model explicitly divides the globe into an empirically defensible North and South. Third, this model does not address a wide range of issues. It uses the trade and development question from chapter 4, "How does the North push off its environmental externalities to the South?" to maintain analytical focus.

5.3.1 Model Structure

The structure of the ELP model turns on the three master variables of lateral pressure: population, technology, and resources. Two of these are captured in the simple overshoot and collapse model presented in figure 5.1: Population is self-explanatory, and Resources is an abstraction of Food, the fundamental and limiting commodity of the overshoot and collapse system. Technology is missing in the overshoot and collapse model, but it is included in the ELP model of figure 5.4 through the proxy variable GNP (Choucri and North 1993b).[48] GNP serves as an intermediary or intervening variable between Population and Resources, all of which are represented by stocks. This three-tiered structure makes intuitive sense because populations interact with the natural environment through their technology.

Population, as with the overshoot and collapse model, is modified by two flows, deaths and births. The deaths flow is driven by the graphical converter mortality, which is determined by GNP per capita. Its graphical function is described more fully in appendix F. The GNP stock is modified by two flows, investment and depreciation, and those too are explored more fully in appendix F. The Resources stock is modified by the consumption and regeneration flows. The regeneration flow represents the rate at which natural resources regenerate themselves, which can best be envisioned in terms of the study's empirical measure, forest area. If a patch of forest is chopped down, it recovers or regenerates at a rate measured in terms of forest area/year. The regeneration flow is actually measured in terms of resources/year. The consumption flow represents the rate at

138   Chapter 5

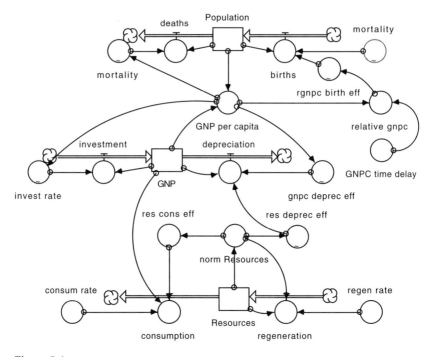

Figure 5.4
Environmental Lateral Pressure (ELP) System Structure. The ELP system dynamics model features three stocks that represent the three master variables of lateral pressure: *Population*, *GNP* (a proxy for technology), and *Resources* (as measured by forest area). Two attendant flows, one contributing to and the other detracting from, are attached to each stock: Population has *births* and *deaths*, GNP has *investment* and *depreciation*, and Resources has *regeneration* and *consumption*.

which natural resources are consumed, a process driven by GNP. Note that the flow aspect of GNP, measured in terms of $/year, is made explicit as a growing GNP requires an increased flow of natural resources to support that growth.

Chapters 3 and 4 make an explicit distinction between the developed North and developing South, and this distinction is carried forth into the ELP model by replicating the structure of figure 5.4 into two distinct regions representing North and South respectively. In figure A.4, it is shown that differential equation models can generate different dynamics given different initial conditions. This same technique is used here to represent North and South with the same system structure: initial Population for the North is 1000 population units and for the South, 2,000 units; initial GNP for both North and South is $2.5 million/year yielding initial GNP per capita values of is $2,500/person-year for the North and $1,250 for the South; initial Resources for the North is 5×10^9 resource units and for the South, 15×10^9. These initial values do not strictly correspond to real-world values but are instead reasonable starting points gleaned from the time-series analysis of section 3.2. The unique dynamics generated by the different initial conditions are presented in section 5.3.3. The two regions, North and South, are then connected through resource and technical trade relationships. Figure 5.5 shows the resource trade

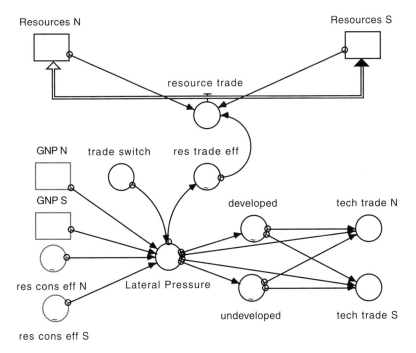

Figure 5.5
Resource Trade System Structure. Because the model must capture trade relations between the North and South, the ELP structure is duplicated with one instance representing the North and the other the South. An appended "N" for North and "S" for South differentiates the duplicated variables. Lateral pressure, essentially the difference between northern and southern GNP, drives resource trade between the regions.

connections. The flow resource trade connects the Resources stocks of North and South, thereby allowing natural resources or raw materials to flow between these geographically distinct regions.

High-technology trade, of the type that generally flows from North to South, is treated somewhat differently than the resource-based trade that typically flows from South to North. Technology-based trade is not quite as simple as resource-based trade for the following reason: if a technology is transferred from the North to the South, then that technology is still available for use in the North. This observation is based on the fact that information can be replicated almost costlessly, physical resources cannot. For example, if the United States shares some software with a developing nation, then the United States retains its ability to use that software. Conversely, if the developing country cuts down a forest and sells its trees to the United States to pay for the software—the forest is gone, at least until new trees slowly grow to replace the old ones. The dynamics of technical trade are captured by the graphs in figure F.6, which yield growth rates that are added to the investment flows of the North and South as shown in figure 5.6. Per the tenets of comparative advantage, both developed and developing countries benefit from trade. In fact, developing countries benefit about five times more than developed countries. These graphs are scaled this way because developing nations tend to have smaller GNPs,

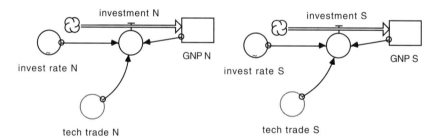

Figure 5.6
Technical Trade System Structure. Technical trade is different from resource trade in that if a high-tech product is shipped abroad, the technology remains. Resources, in contrast, when shipped are no longer available. Thus, the technical trade model involves investment contributions for North and South (cf. figure F.6). Resource trade (cf. figure 5.5) depicts a material transfer between North and South.

and so an infusion of high technology goods and information will generate a larger percentage impact. Conversely, developed nations tend to have large GNPs, and so the additional profits from technical trade tend to pack a somewhat smaller economic punch.

5.3.2 Model Feedbacks

The analysis of the ELP model's feedback characteristics comes between the structural and dynamic analyses because feedback helps one make sense of the transition from a system's microfeatures to its macrobehavior. The web of causal connections specified in the system dynamics model forms feedbacks that determine the ELP model's dynamic response. Because the relationship between system structure and dynamics is complex and nonintuitive, building a workable model allows for inquiry, experimentation, surprise, and learning. In keeping with the previous section's organization, the intraregional feedback relationships of figure 5.4 are developed first. This portion of the analysis will focus on lateral pressure's three master variables: population, technology, and resources. Second, the interregional trade relationships of figures 5.5 and 5.6 will be examined. Here the primary focus will be on how trade ties together North and South.

Figure 5.7 shows the feedbacks that pertain wholly to the social environment. The central variable, GNP per capita, connects the two master variables, Population and technology as represented by GNP. The feedbacks surrounding Population are similar to those presented in

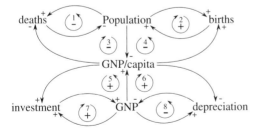

Figure 5.7
Population and GNP Feedbacks. The system structure between the *Population* and *GNP* stocks form eight feedbacks. The positive feedbacks tend to dominate causing Population and GNP to increase together.

Figure 5.2 in that a negative feedback results from Population's connection to deaths (loop #1) and a positive feedback from its connection to births (loop #2). A similar set of relationships surround GNP with a positive feedback resulting from the investment connections (loop #7): more GNP leads to more investments and more GNP, creating the self-reinforcing relationship characteristic of positive feedback. Conversely, GNP's connection to depreciation yields a negative feedback (loop #8): more GNP leads to more depreciation and less GNP, creating the self-abnegating, balancing relationship characteristic of negative feedback.

GNP per capita influences all four of these feedback relationships. Loop #3 connects GNP per capita, deaths, and Population resulting in negative feedback. This implies an increase in GNP per capita tends to decrease deaths, increase Population, and decrease GNP per capita. The same holds true for loop #4 in which GNP per capita increases births, thereby increasing Population and decreasing GNP per capita. Increases in living standards therefore contribute to population growth which undercuts these same gains, both in terms of decreased deaths and increased births. These negative, population-related feedbacks provide contrast to the two positive, GNP-related feedbacks. Loop #5 connects GNP per capita, investment, and GNP in a self-reinforcing, positive relationship—an increase in any single variable leads to increases in the other two. Loop #6 is a little more complicated: an increase in GNP per capita leads to a decrease in depreciation, resulting in increased GNP and GNP per capita. Given these relationships, it is easy to see how a growing economy creates the conditions for its continued growth and how population increases can undercut this growth.

Figure 5.8 shows the feedbacks that pertain to the divide between the social and natural environments with GNP representing the social and Resources representing the natural. Note that feedback loops #7 and #8 are the same as those in figure 5.7. Moreover, the feedbacks surrounding Resources are similar to those of Population and GNP. Loop #10 combines Resources and consumption creating negative feedback: more resources leads to more wasteful behavior and consumption, which reduces resources. Conversely, fewer resources lead to more efficient behaviors and reduced consumption, which conserves resources. Loop #11 is a positive feedback between Resources and regeneration. This loop is best

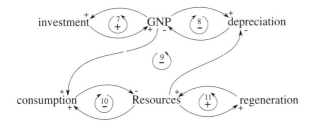

Figure 5.8
GNP and Resource Feedbacks. The system structure between the *GNP* and *Resources* stocks form five feedbacks. The central negative feedback, loop #9, dominates thereby depleting Resources and limiting GNP.

envisioned in terms of this study's empirical referent, forest area: more trees provide more seeds for even more trees. In this fashion, the natural environment provides the means for its continuation (Lofdahl 1992, 65). Negative feedback loop #9 is perhaps the most important because it connects GNP with Resources and in so doing explicitly connects the social and natural environments. Essentially this loop shows that increased GNP leads to increased consumption and reduced Resources. Continuing around, shortages in Resources lead to increased depreciation that eats into GNP. Therefore, to maintain continued GNP growth, Resources cannot be allowed to drop too low. This requires not surpassing the regeneration rate or acquiring additional resources from elsewhere through trade.

Figure 5.9 shows the feedbacks that result from establishing trade relationships between North and South.[49] Loops #12 and #13, which connect GNP and Resources for North and South respectively, essentially match loop #9 in figure 5.8. Central to this diagram is *Lateral Pressure*, which controls *technical trade* from North to South and *resource trade* from South to North. Environmental lateral pressure is determined by differences between GNP.N and GNP.S, with the stronger economy expanding into the weaker. The developed North has historically expanded into the relatively undeveloped South, and the discussion proceeds taking this empirical fact into account. Loop #14 shows that economic growth in the North increases Lateral Pressure, which increases technical trade and GNP.N. Conversely (loop #15), technical trade increases GNP.S, which helps the South to achieve economic parity and mitigate Lateral

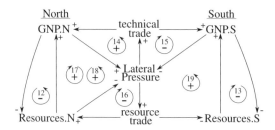

Figure 5.9
Trade Feedbacks. The trade relationship between North and South forms these feedbacks. Note that increasing northern GNP also increases lateral pressure, technical trade, and resource trade.

Pressure. Note that *technical trade* increases both GNP.N and GNP.S per the tenets of comparative advantage. On the other hand, resource trade increases Resources.N and decreases Resources.S, thereby creating loop #16. Increasing Resources.N lessens Lateral Pressure, but this effect tends to be small compared to the driving GNP differences.

The final three feedbacks—positive loops #17, #18, and #19—help the most in understanding the upcoming dynamic analysis. Loop #17 connects GNP.N, Lateral Pressure, resource trade, and Resources.N. Every connection in loop #17 is positive, creating a self-reinforcing, positive feedback. Increases in any aspect of this loop amplify themselves throughout the remainder of the loop. Loop #18 goes counterclockwise connecting GNP.N, Resources.N, Lateral Pressure, and technical trade. This loop contains two negative causal connections: As GNP.N grows, Resources.N become increasingly depleted, which increases Lateral Pressure. This in turn leads to increased technical trade and further GNP.N growth. Note positive feedbacks exist for the developed North through both technical trade and resource trade. The final feedback, loop #19, winds its way through GNP.S, Lateral Pressure, resource trade, and Resources.S. This loop can best be understood by starting with declining Resources.S, which lowers GNP.S thereby increasing Lateral Pressure and resource trade causing even more resources to be exported to the North. This too is a self-reinforcing process, but one that leads the South toward economic stagnation, not growth as in the North. Such dynamics are developed and analyzed more completely in the following section.

5.3.3 Model Dynamics

This section demonstrates how the complex, causal relationships defined through system structures and feedbacks play out over time. This is accomplished by following the trajectories of lateral pressure's master variables—population, technology, and resources—over two scenarios. First, the North is examined as if it had no trade connections at all with the South. This establishes an analytical baseline from which more realistic scenarios can be evaluated. Second, trade connections between North and South are inserted into the ELP model by turning on the trade switch (cf. figure 5.5). The changes that result from the model with trade "turned on" can then be examined in light of the previous scenario without trade.

Figure 5.10 presents the behavior of the ELP model for the North without trade connections to the South. Lateral pressure's three master variables are presented: *normalized Population.S, normalized GNP.N,* and *normalized Resources.S.* "Normalization" implies dividing a variable by its initial value to simplify the analysis by giving an instant idea how the

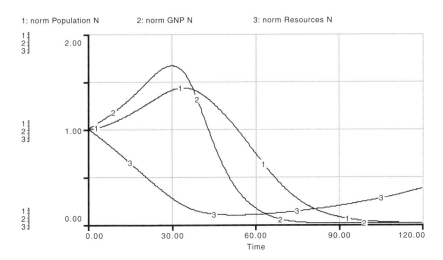

Figure 5.10
North Dynamics without Trade. *Population* and *GNP* increase so long as *Resources* hold out. When Resources become sufficiently depleted, Population and GNP fall in a manner reminiscent of the overshoot and collapse example (cf. figure 5.3). Note that all values are *normalized* by dividing through with their initial values. Thus all three variables start from 1 at time 0.

variable has changed since time 0 (when all normalized variables start at 1). For example, if over the course of the simulation a normalized variable—say, Population.N—equals 2, then one knows that the variable has doubled from its initial value. The simulation lasts for 120 years, and it helps to consider it starting at the end of World War I or 1920, a point of hegemonic transition from Great Britain to the United States. However, the initial values do not correspond to those of 1920. The initial values constitute reasonable starting points based on the contextual analysis of chapter 3. What matters most is not the actual values, but the relationships among the variables, and so rough, order-of-magnitude values are sufficient so long as their relative relationships are representative and consistent.

The dynamic response of figure 5.10 shows GNP.N growing rapidly, which causes Population.N to grow and Resources.N to fall.[50] The primary cause for economic growth comes in the relative difference between Population.N and GNP.N: because the North starts off with half the population and the same GNP as the south, it has twice the GNP per capita which allows for savings, investment, and capital accumulation. However, the growth of GNP.N can only be maintained for thirty years until Resources.N gives out. From year 30 onward, GNP.N falls as does Population.N until they both collapse, at which point Resources.N begins to recover. The South without trade (cf. appendix G), in contrast, exhibits a stable and sustainable equilibrium without the wild, dynamic swings experienced by the North. The scenario here is similar to that proposed by Malthus and explained by overshoot and collapse: economic growth can only be maintained so long as undeveloped land provides new natural resources that sustain economic growth. Faced with this problem, Malthus proposed accepting limited resources, embracing morality, and having children only if their material support could be assured. Ricardo proposed a different solution, trade.

The effect of trade on Northern economic growth, as portrayed by figure 5.11, is striking. Trade triples GNP.N's time of growth from thirty years to almost ninety. Moreover, GNP.N climbs to almost five times its initial value, necessitating a new graphical scale to accommodate the curve. The growth of GNP.N easily outstrips the growth of Population.N, which levels off in the face of increasing GNP.N between years 60 and

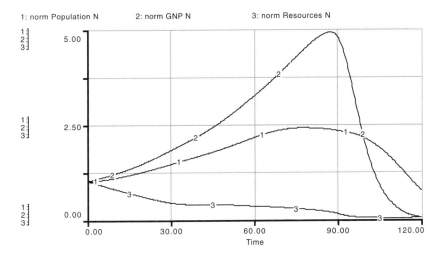

Figure 5.11
North Dynamics with Trade. Again, *Population* and *GNP* increase so long as *Resources* hold out. However, with trade granting access to southern Resources, northern Population and GNP grow much larger over a longer period. Note also that GNP grows much larger than Population, leading to an increase in GNP per capita. When Resources run out, Population and GNP collapse as before.

90. These responses closely parallel the time-series results offered in section 3.2, although a longer time-frame is covered here. International trade solves the problem of land and resource constraints for a time, just as Ricardo argued, as trade provides access to southern resources so that the North can maintain its economic growth. After year 90 though, GNP.N and Population.N collapse just as they did previously. More curious though is the response of Resources.N. Why does the curve level out between years 30 and 90, just when GNP.N experiences its most dramatic growth? After all, it was the collapse of Resources.N that retarded economic growth in figure 5.10. How does international trade provide both superior economic growth and comparative resource stability in the North?

Figure 5.12 shows that the South, per the tenets of comparative advantage, experiences economic growth as does the North. Just as the South provides resources to the North, the North provides technology to the South, and both partners benefit. However, the growth of Population.S

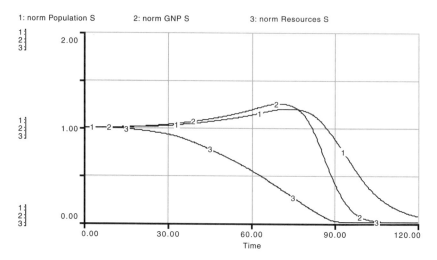

Figure 5.12
South Dynamics with Trade. With trade, the South experiences both *Population* and *GNP* growth. Although unlike the North, Population increases are sufficiently large and GNP gains sufficiently small to leave the South's GNP per capita essentially constant. Depletion of southern *Resources* however is much more pronounced, and when they run out, Population and GNP collapse.

follows closely behind that of GNP.S, and so the South never experiences commensurate gains in GNP per capita. Also, Resources.S experiences steady depletion from year 30 until it is completely exhausted in year 90. These behaviors of growing southern population and GNP, and failing resources closely parallel the time-series data presented in section 3.2. These observations can be explained in several ways. First, the period of rapid southern resource decline, between years thirty and ninety, coincides with the North's period of surprising resource stability. This occurs through the North's importation of the South's raw materials in the form of traded commodities, which effectively pushes off the environmental costs of the North's economic growth to the South. Second, the South experiences modest economic growth compared to the North, despite the fact that technical trade contributes much more significantly to the economic growth of developing countries than developed (cf. figure F.6). The ELP model therefore helps to explain how the "small tail" of international trade wags the large, developed economy dog (Samuelson and Nordhaus 1992, 487): Trade's direct economic contribution does not drive northern growth so much as it prevents resource shortages from physically limiting it.

A third observation, which helps explain the economic divergence between North and South, comes in comparing the northern dynamics without trade (figure 5.10) against southern dynamics with trade (figure 5.12). The two figures both exhibit an initial period of economic and population growth accompanied by resource depletion, followed by subsequent population, GNP, and resource collapse. Two things separate the scenarios. First, the northern response is driven by and confined to the North—its economic and resource collapse does not affect the South (cf. figure G.1). The southern collapse is driven primarily by northern lateral pressure, which *synchronizes* northern and southern economic and environmental collapse in year 90. Because southern resource utilization is driven by the North, this explains how southern environmental degradation takes on a similar, synchronized character across vastly different cultures, economies, and ecologies (Panayotou 1992, 317). Another aspect that separates the isolated, northern collapse scenario from the coordinated, southern collapse is the time it takes to achieve resource exhaustion—thirty years in the North, and ninety years in the South. The lengthened time-frame is directly attributable to the greater geographical area and resources of southern countries.[51] The northern collapse is regional in scale, and natural resources soon recover because they are not completely exhausted. The southern scenario is instead global in scale, and because global resource depletion is driven down that much further by the additional trade feedbacks of figure 5.9, global resources are depleted much more thoroughly and, consequently, find it harder to recover.

5.4 Critical Inferences and Implications

The ELP model developed throughout this chapter was created to evaluate, interpret, and connect three analyses: (1) the time-series analysis of chapter 3, (2) the statistical analysis of chapter 4, and (3) the economic analysis of comparative advantage and overshoot and collapse. The ELP model should not be evaluated by whether its dynamic outputs forecast "the truth" but rather on its ability to synthesize the various relevant theories, relationships, and data into a plausible explanation based on

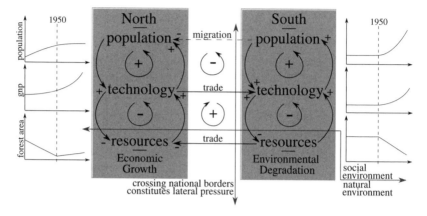

Figure 5.13
Environmental Lateral Pressure (reprise). This diagram, originally presented as figure 2.4, is shown here to reprise the major themes developed in this chapter. First, the ELP model's structure is based on the three master variables of lateral pressure—population, technology, and resources—for both the North and South. Second, these three variables are connected so as to form multiple feedback loops. Third, the model generates dynamic responses consistent with the graphs to the left and right.

a coherent model that generates illuminating scenarios. The ELP model does this not by addressing a wide range of issues but by connecting the theoretical and empirical components of this study by tackling the question "How specifically does the North push off its environmental costs to the South?"

First, the time-series analysis centers on figure 5.13, which recalls the original conception of environmental lateral pressure as presented in chapter 2. One measure of success for the ELP model rests in the level of fidelity its dynamics achieve with the hypothetical response portrayed to the left and right of figure 5.13, both theoretically and empirically. Theoretically, the model must be judged in terms of lateral pressure theory and how well its representation of environmental lateral pressure adheres to the theory's basic tenets. Empirically, it will be judged by how well it explains the figure's hypothetical dynamics that summarize those presented in section 3.2. This concluding discussion will be organized along the same lines as this chapter: structure, feedback, and dynamic response.

The ELP model adheres closely to the tenets of traditional lateral pressure theory by grounding its structure in the three master variables: population, technology (as measured by GNP), and resources (as measured by forest area). Additionally, the very concept of lateral pressure is used to drive the technical and resource trade processes that transcend national borders thereby connecting geographically separate regions. Another analytically useful aspect of the ELP model is the way it treats GNP as a flow. Beyond measuring GNP in terms of dollars per year ($/year), it directly drives the resource consumption flow (cf. figure 5.4). In this fashion, the relationship between GNP and resource consumption is explicitly developed. Less satisfying is the way GNP per capita drives the dynamics of Population and GNP. Certainly there is more behind political development and social capital than changes in GNP per capita, and this study explicitly argues that GNP per capita does not accurately measure living standards. So long as GNP/capita is used as a measure of development and not of true living standards, then it remains a useful and defensible measure. Finally, it is recognized that Resources aggregates and simplifies the complex processes of the natural environment, especially as regards resource regeneration. A commodity like oil should have a large stock and a small rate of regeneration, while forest area might have a smaller stock and a larger regeneration rate. Given this study's focus on forest area and the additional complexity that a more accurate portrayal would require, this simplification is also deemed useful and defensible.

Figure 5.13 contains three major feedbacks: (1) the positive loop between population and technology, (2) the negative loop between technology and resources, and (3) the positive loop between technology and resource trade.[52] Each is disaggregated and developed further in this chapter's feedback analysis section. This exercise does two things: It dives into the microfeatures that underlie these high-level feedbacks per section 1.3 (and appendix A), and it helps in the transition from a static, structural analysis to a dynamic one. First, the positive loop between population and technology is examined more fully within figure 5.7. Here it can be seen that the positive loop that connects population and GNP actually consists of four feedbacks, two positive and two negative. Should the positive loops dominate, then capital accumulation and economic growth result. Should the negative loops dominate, then population growth undercuts economic gains. Second, the negative loop between technology and resources is examined within figure 5.8. The negative feedback between GNP and resources is unambiguous and, constitutes the most explicit and major linkage between the social and natural environments. Third, the positive loop between technology and resource trade is developed within figure 5.9. This positive loop, when examined in more detail, becomes three positive feedbacks, two of which help along economic growth in the North (loops #17 and #18), and the other (loop #19) contributes to economic stagnation in the South. These trade relationships also link the social and natural environments.

The dynamics of North and South correspond well to those presented in figure 5.13 and more importantly to those in section 3.2, although it must be remembered that data exists for only fifteen to twenty-eight years and the ELP model allows for much longer scenarios. The model was able to recreate northern economic growth and southern stagnation and environmental degradation. Two explanations fall out of the analysis that explain chronic southern underdevelopment: The local variant is that southern GNP per capita never rises high enough to allow for capital accumulation, and the global variant is that southern resource depletion contributes to its low economic growth. The ELP model also allows for various "what if" analyses such as "What if the North and South were not connected through trade?" This analysis reveals an environmentally stable South, indicating that economic collapse is not "dialed into" the model—sustainable systems are achievable. Additionally, the collapse of the North without trade is isolated and regional, and the collapse of the North and South with trade is synchronized and global. Finally, the trade scenario shows that the hope of all countries enjoying a northern standard of living is chimerical as the North's consumption is predicated on asymmetric trade relationships with the South. Were the South to live similarly, with whom would they trade and from where would they garner their additional resources?

These observations lead to the second goal of this analysis: explaining the statistical results relating trade and the environment. Whereas previous analyses have reported a correlation between high GNP and a healthy environment, this study shows that trading with high GNP partners leads to domestic deforestation. This model demonstrates how a divergence between northern high-technology and southern resource endowments yields a South to North resource transfer per the tenets of comparative advantage. The fidelity of the system's dynamic response to history provides additional evidence that high GNP growth drives lateral pressure through the logic of comparative advantage, which grants rich countries access to the resources of the poor. Of course the model is simplified: Countries are more varied than just North and South, trade is more varied than just technology for resources, and resources are more varied than just trees. Nevertheless, comparative advantage works on all country pairs at various gradations of northness and southness, wealth and poverty, and the sum of these myriad trade interactions, and many other interactions besides, results in economic growth for the North, environmental degradation for the South, and globalization for all.

Thus the third goal of the model, addressing the trade and environment debate in terms of its creators, Ricardo and Malthus, has been achieved. They are reconciled insofar as they both characterize key aspects of a larger, more complex international trading system. Comparative advantage has been shown to increase GNP in both North and South, although the South experiences the greatest share of environmental degradation, and population gains cut into southern GNP per capita. The decline or down side of the overshoot and collapse dynamic put forth by Malthus and *The Limits to Growth*, in contrast, is absent from the historical record. Any potential sharp reductions in population or economic output remain safely in the future. But the blending of comparative advantage with overshoot and collapse using environmental lateral pressure and system dynamics has revealed that the implicit logic of comparative advantage increases resource extraction and, thus, increases the strain on the natural environment. After all, the additional goods people consume through comparative advantage must come from somewhere. That the extra bolts of cloth and bushels of wheat in table 5.1 entail an environmental cost is observable even without a supporting computer simulation. Nevertheless, the simulation shows that comparative advantage and trade do increase resource extraction, exacerbate global environmental degradation, and contribute to the pervasive resource depletions necessary for overshoot and collapse to occur. Some have argued that because the mineral and oil shortages predicted by *Limits* did not actually occur, the logic of overshoot and collapse is flawed (Economist 1997a). As has been demonstrated throughout the twentieth century, the global economy and its trading system are excellent at delivering technologies and product substitutes that prevent material shortages from limiting population and economic growth. However, current population and environmental trends may soon lead to shortages not of minerals and oil but of food and fresh water, key commodities for which suitable substitutes are hard to find (Litvin 1998a). Without significant policy changes in the twenty-first century, Malthus may yet have his day.

6
Conclusion: Implications for Theory, Methods, and Policy

In the aftermath of the failed 1999 WTO meeting in Seattle, concerned observers are still trying to figure out just what happened. Seattle was to kick off a new round of trade liberalization talks that would lower tariffs, increase trade, and deliver economic growth. There was little reason to believe anything would go wrong as the WTO was simply continuing the process of globalization that had been going on for over fifty years. But something did go wrong. Thousands of antiglobalization protesters including anarchists, labor activists, and environmentalists disrupted the meeting. When the WTO delegates did finally get together, they failed to agree on an agenda for a new round of trade talks. The basic problem was that delegates from the developed North, who were the most concerned with labor and environmental standards, failed to find common ground with disgruntled delegates from the developing South, who were more interested in increased economic growth and access to rich, northern markets.

One question about Seattle regards the reason for its failure to deliver another round of trade talks—was it due to the protests or to the disagreements among delegates? The answer, based on the analysis presented in this study, is yes. That is, the very terms of trade between North and South generate the costs that concern both the protesters and the delegates, thus, determining who is more responsible is less important than understanding the root causes of the problems and figuring out what to do about them. Instead of drawing semantic distinctions, this study presents a more systemic analysis that investigates causal interconnections among key global measures. Specifically, it is found that the structure and terms of international trade lead to the environmental degradation that is of concern to the North as well as the economic stagnation that is of concern to the South.

That trade is bad for economic development and the global environment is a controversial claim. Economically, the theory of comparative advantage (cf. table 5.1) holds that any country that participates in international trade, either rich or poor, benefits in the form of increased consumption. The comparative advantage argument is accepted, and its assumptions are built into the Environmental Lateral Pressure (ELP) model (cf. section 5.3). Indeed, the simulation shows that Gross National Product (GNP) grows under the influence of trade both in the North and South. However, while the North experiences increased GNP per capita, southern population growth undercuts these gains, leaving the region relatively stagnant economically. These somewhat theoretical claims are empirically supported by the time-series analysis found in section 3.2. So while the theory of comparative advantage is true—trade contributes to GNP increases in both regions—complex relations with other key variables and processes work against realizing these economic gains (i.e., Southern population growth).

It has been asked whether trade helps or hurts the global environment (cf. chapter 4). Just before Seattle, *The Economist* (1999a) explained "Why Greens Should Love Trade." The article stated that economic studies show that rich countries have the best environments. The reason for this relationship is less clear than the fact it exists, but economists argue that rich nations have better environments because they have the resources to afford them. Poor countries, in contrast, tend to have worse natural environments, which prompts economists to conclude that if all nations were rich, then the global environment would be in better shape. Environmentalists argue in response that the natural environment was doing pretty well until industrialization came along, and that the continued spread of international development through globalization and trade will only further degrade the natural environment.

This study explicitly tests whether trade helps or hurts the environment by employing a newly developed empirical measure, Trade Connected GNP (TC × GNP), which represents the aggregate GNP of a country's trading partners. TC × GNP is tested along with GNP per capita and population growth to explain this study's measure of environmental health, deforestation or forest change. When these three variables are used to explain deforestation, GNP per capita and population growth perform as expected. As economists contend, high GNP per capita means more forests, and low GNP per capita means less. Population growth, as expected, is inversely correlated: High population growth means less forest, and low population growth means more. The surprise comes with TC × GNP: Like population growth, increased TC × GNP means less forest, low TC × GNP means more. However, testing reveals that TC × GNP yields more explanatory power than the two more traditional environmental economic measures, GNP per capita and population growth (cf. section 4.4 and appendix D).

The statistical results generated by TC × GNP provide an opportunity for interpretation. Rather than maintain that trade helps the environment as economists do, this study argues that *trade provides the mechanism by which the costs of industrialization are pushed off by rich countries onto poor ones.* This conclusion is supported by the system dynamics analysis of chapter 5. The ELP model depicts the developed North exporting high technology, high value goods to the developing South, while the South exports natural resources to the North. These terms of trade deliver asymmetric benefits: economic development and a clean environment for the North; economic stagnation and environmental degradation for the South. This result is supported by the widely divergent globalization literature reviewed in section 1.1. If one works for a rich, northern MNC and is reaping the rewards of globalization, then one will tend to support trade. If one is instead an unemployed citizen of a poor, southern country, then one is more likely to be suspicious of globalization's benefits. That wealth drives environmental externalization also makes sense from a micro perspective: if the citizens of a rich, northern country desire a product with prohibitive environmental costs—say, cheap wood furniture—then if given the opportunity, they will import the product from a developing country rather than pay higher prices, or they will do without. Such decisions are made almost automatically by the market's, "invisible hand," made all the more invisible by the modern, international economy. Developing countries, in some sense, are forced by their poverty into responding to such forces and selling their natural resources, thereby placing them into the also-ran role of commodity provider.

The WTO demonstrations in Seattle, in light of this analysis, can be viewed as an intellectual wake-up call. Globalization is creating relationships of such complexity that traditional theories and debates prove insufficiently illuminating. The obviousness of global environmental degradation coupled with a partially responsible institution that was slow to address the topic eventually led to the frustrations, demonstrations, and riots of the anti-WTO protests. However, it is doubtful that the Seattle protests alone will generate a conceptually richer and more mutually beneficial policy dialogue. To help bring this about, this study has addressed a coterie of theoretical and methodological issues that are reviewed in the next two sections. These issues, in turn, have policy consequences that are discussed in the third section. In the final conclusion several next steps are suggested that would continue and expand the research agenda presented in this study.

6.1 Theoretical Extensions

From a complex systems perspective, the political and economic theories that led to the creation of today's international system are fairly simple. This is not a pejorative observation: given the paucity of computational capacity available at the time the foundation of the world economy was established, that such a workable, successful, and large-scale system could be created at all is a testament to the creativity, dedication, and intelligence of those involved. The work was started at Bretton Woods, New Hampshire in 1944 where two major post–World War II international institutions were created, the World Bank and the IMF. The General Agreement on Tariffs and Trade (GATT) joined them later in 1947. Together these institutions, informed by modern economics, have lowered tariffs, invested capital, and spurred economic growth around the world. The trade, technology, and wealth so evident today in the developed world are direct results of these postwar economic efforts.

A good portion of this economy building and globalization effort was expended to prevent war, the traditional concern of international relations. Prewar tariffs led to worldwide depression that helped radicalize Germany and bring about National-Socialism. This relationship between tariffs and war helps to explain the zeal with which today's economists seek tariff reduction and trade promotion. Trade and the growth it brings are historically tied to notions of social progress and furthering the general welfare. Chapter 2 shows that economics, in addition to preventing war, contributes methods to study it. The appeal of economics derives from its aforementioned success and analytic rigor. It is arguably the premiere social science from a methodological perspective. It is natural to progress from initial theoretical articulation to subsequent methodological implementation, and economics provides an established, social science path to do so.

Waltz (1959) offers just such an initial theoretical articulation in his three images for the study of war: the individual, the state, and the international system. Waltz (1979) then incorporates economics as the analytic means to inform his power-based study of war. Keohane (1986a) critiques this effort, noting well the analytic limitations of the methods employed. The primary problem was that the representations of the individual, state, and international system were insufficiently rich to achieve its stated goal, that of dynamic description. Waltz, however, has written extensively on the analytic richness required to study of war; the problem was not his intuition but his methodology. Economics was too analytically constrained and static for the job.

Choucri and North (1975) also seek rigor when using *lateral pressure* to study war, but they do so by partaking from a different analytic tradition—that of geopolitics, geography, and the natural sciences. Their effort begins by breaking Waltz's single explanatory variable, power, into three: population, technology, and resources. Employing a multivariate analytical foundation implies causal connections among the variables of interest. In Choucri and North (1975) and Choucri, North, and Yamakage (1992), lateral pressure's multivariate structure is used to study the causes of war, although it is extensible to other topics as well. Upon articulating this analytic division—that is, splitting a single explanatory variable into three—lateral pressure becomes a *linkage* theory that employs the dynamic language of the natural sciences, differential equations. In so doing, Choucri and North bridge the natural and social environments by combining scholarship from the social and natural sciences. Specifically, lateral pressure uses analytic techniques originally developed in the natural sciences to study social science questions.

Choucri (1993a) applies lateral pressure theory to the study of global environmental degradation, specifically investigating carbon dioxide effluents. Since lateral pressure is more of a perspective than a specific topic, its application to a different though related area of study is both understandable and defensible. From an abstract perspective though, Choucri (1993a) makes even more explicit the conceptual connections between the social and natural environments. In identifying and studying the anthropocentric causes of global environment degradation, this study maintains that processes from within the social environment cause problems in the natural. Once again, this denotes a linkage challenge, but how best to study it? A range of methodologies are drawn from the natural sciences and placed into the service of social science scholarship. They are reviewed in section 6.2.

In acknowledging the conceptual and causal connections between the social and natural environments, this study finds it useful to review the motivation and stakes of the driving question. Why be so concerned with global environmental degradation in the first place? Why make the effort? Although environmental degradation has been a part of civilization as long as there has been civilization, never before has synchronized environmental degradation taken place on such a scale that it threatens the total, global environment. The stakes of global environmental degradation are high as it is impossible for individuals, nations, and the international system to exist without a sufficiently nurturing global system in which to live (North 1990). The question of global environmental degradation thus represents a conceptual sea change. Until now, the social environment has concerned itself with expanding into the natural environment. To continue doing so threatens the natural environment, which in turn threatens the social. The continued existence of the social environment thus depends on adequately accounting and caring for the natural environment. Paradoxically, looking outside the social environment to address, understand, and incorporate the natural is imperative because it is the social environment that is both being threatened and causing the problem.

Finally, in addressing the relationship between humanity and nature, here called the social and natural environments, it is realized that these are venerable topics, about which much has been written. However, just as advanced technology has given society the potency to threaten the natural environment, so too has technology granted the means to study it in new ways and from fresh perspectives. Global environmental degradation is a complex issue. There are myriad relevant systems including the economic, political, cultural, and ecological, each of which are sufficiently detail laden to merit significant study on their own. Combining them only increases the complexity. However, recent advances in computation and the study of complex natural systems have yielded analytical techniques that can be used to study complex social systems, systems like the international, trade-based economy. So while questions of humanity and nature are in some ways classic, the scale of the problem and the power of the tools available to study them are quite new. The specific methodologies used to study global-scale systems and their attendant complexities, as organized by the theory of environmental lateral pressure, are described in the following section.

6.2 Methodological Advances

The goal of richer, more encompassing, and more descriptive theories has long been desired (cf. section 1.2.2), and recent advances in computation have made this goal more attainable. At the most basic level, computers handle the many niggling details implied by complex theories. Simple theories describe the relationship among only a few key variables. Comparative advantage, for example, relates production, consumption, and trade for two countries. Complex systems entail many more variables and relationships that would quickly overwhelm any person regardless of intelligence, but a computer easily accommodates and effortlessly tracks many more causal relationships than can a person. Specifically, a properly directed computer can greatly facilitate exploring a complex problem like trade and the environment. With the computer handling the system's details, human cognition is then freed to do what it does best, recognize relationships and match patterns. Often the key element or relationship of a complex system remains locked away, hidden within the empirical data because it has not been displayed in a way that can be interpreted, recognized, or even understood (Gleick 1987; Stewart 1989; Hall 1991). The three methodologies used in this study—GIS, statistical and diffusion models, and system dynamics—together display and explain the empirical data so as to reveal the complex causal connections that link the social and natural environments.

The empirical data is first encountered through GIS–produced maps, which yield a methodological "snapshot." Although data for only a single variable and a single timeperiod can be presented, spatial patterns become readily apparent. For example, the maps presented in section 3.1 show that high GNP and CO_2 countries are in the north and high population and deforestation countries in the south. While this observation is not entirely new, seeing the geographic relationships displayed in graphical form allows for a more intuitive understanding that helps spur new questions and experiments that might, in turn, reveal deep and surprising revelations. GIS is used here to introduce the study's issues, concerns, and variables. Without reaching any final conclusions, the geographic analysis establishes the context of the study.

The multivariate statistical analysis is broken up into two sections, time-series and pure statistical analyses. The time-series analysis provides explicit contrast to the geographical analysis in that the graphs of section 3.2 demonstrate change over time but no explicit spatial distribution. So, GIS and time-series analysis together provide a complementary way to perform exploratory data analysis as together they address both spatial and temporal variation. The time-series displays are geographically disaggregated into North and South to give a sense of how the measures change in different parts of the world. Here it is observed and established that GNP, trade, and deforestation are all growing trends. This naturally leads one to question whether they might be causally related.

Within the pure statistical analysis, correlations between trade and environment were tested to provide insight into potential causal connections. The primary difficulty was finding some way to represent and test the effects of trade on the environment. Testing trade is hard because of its systemic complexity. Comparative advantage is typically explained using only two countries, but the real trading system consists of hundreds of countries that trade with each other. Trade-based contiguity matrices were used to create a variable, Trade Connected GNP or TC × GNP, that explicitly represents and accounts for trade connections among countries. In this fashion, the resulting diffusion model captures the hyper-connectedness implied by globalization. TC × GNP shows that GNP increases in one country lead to deforestation not in that country *but in its poorer trading partners*.

As provocative and compelling as the TC × GNP results might be, correlation is not causation. Thus the results might simply be a statistical artifact, so a clearer causal explanation was developed. This task was accomplished using a system dynamics simulation model that complements the statistical analysis in much the same way GIS complemented the time-series analysis. Whereas the statistical model manipulates data, captures the hyper-connectedness of complex systems, and yields correlations, the system dynamics model manipulates relationships, captures systemic nonlinearity and feedback, and yields causal structure. After following a three-tiered development process of structural, feedback, and dynamic analyses, the resulting model shows how the North extends its economic growth by importing natural resources from the South, which effectively synchronizes natural degradation throughout the developing world. Although the system dynamics model is not definitive, it provides a well-defined and defensible interpretation of the statistical results as well as an explicit representation of the complex causal connections that link the social and natural environments.

Advanced analytical techniques are increasingly recognized as necessary to analyze the complex questions that confront today's social scientists (Axelrod and Cohen 1999; Homer-Dixon 2000). This study applies advanced computational techniques—GIS, diffusion models, and system dynamics—to an important international question. In so doing, the characteristics of complex social systems can be articulated more generally. The use of GIS shows that complex social systems exhibit *spatial variation*. The maps presented in chapters 3 and 4 were not uniform and consistent. Instead, there were pockets of high and low values in sometimes surprising and unexpected patterns. Even if the data are standard and unsurprising, seeing them in a highly visual display as opposed to a numerical or tabular format casts them in a fresh light that invites renewed consideration. The diffusion model concept captures the hyper-connectedness of complex social systems. For example, the TC × GNP variable explicitly represents the trade connections among over 200 countries, which results in over 40,000 separate trade relationships per matrix. These trade relationships are then calculated for each of the fifteen years of the study, resulting in over 600,000 separate calculations. Performing this analysis without the aid of a computer would clearly be impossible.

The system dynamics model (see chapter 5) says perhaps the most about complex social systems, implying that they can be characterized by stock and flow, nonlinear, and feedback relationships. A combination of high-quality visual displays and underlying differential equations are made possible by the computer, which handles the thousands of modeling details that would overwhelm a computationally unaided analyst. The working description of complex social systems developed here—spatially distributed, hyper-connected, and marked by stock and flow, nonlinear, and feedback relationships—may be incomplete, but it is certainly a good start. At this point, it is more important to recognize that this working definition is sufficiently complete to yield the most important characteristic of complex social systems, *unpredictability*. The behavior of even simple social systems is notoriously difficult to predict, which explains why people continue to study them and formulate policies to deal with them. This topic is developed more fully in section 6.3.

6.3 Policy Consequences

Formulating and implementing policies requires working in and thinking about the real world. Policies are created to address serious problems, which makes the consequences of misguided or poorly implemented policies serious as well. The formulation of policy is informed by theories, methods, and empirical data, and it requires measures of both conceptual foresight and actual experience. Before the advent of computers, the limited power of analytic methods and the paucity of empirical data limited the complexity of informing theories. Now that advanced computers have extended the capability of methods and increased the availability and amount of data, the challenge now is in working with more complex theories and crafting more sophisticated policies, both of which are necessary to understand and direct today's more complex social systems. This study's extension of lateral pressure and its creation of the ELP model are offered as examples of complex theory creation. The ELP model was designed to explicate limited aspects of the international system and demonstrate the capability and applicability of system dynamics. The model is essentially theoretical and therefore inappropriate to serve as a basis for policy creation, although system dynamics has a rich policy history (Sterman 2000). This section considers the policy consequences of complexity in the form of lessons learned from system dynamics. These lessons will be applied more directly to policy formation in the final section.

System dynamics models, while not new, are not yet widely implemented nor understood. However, they have been applied to a wide range of policy problems (Forrester 1961, 1969, 1973), and there exists a reasonable body of experience from which to draw general conclusions about the social environment (Forrester 1971, 59). First, social environments are inherently insensitive to most policy changes, which is a function of the multiloop, nonlinear, and feedback nature of the social environment. The feedback structure itself "balances" and resists efforts to ameliorate social ills. Consequently, social policies that may seem obvious from a legal or political perspective rarely prove effective from a historical or systemic perspective. Second, effective policies and levers frequently do exist, but they are often far removed causally from the consequences that motivate their implementation. Thus, effective social policies tend to be hidden, obscure, unexpected, or counterintuitive. Moreover, experience shows that if by chance the proper policy lever is located, then, as likely as not, it will be pushed the wrong direction, thereby exacerbating the problem. Third, a fundamental conflict exists between the short and long-term consequences of social policies. Changes that bring about short-term benefits often come at the expense of long-term sustainability, and changes that remain viable in the long-term may seem irrational or unreasonably painful in the short.[53]

The challenge at hand is to apply these general observations so that they encompasses international trade and the global environment. This process begins by characterizing the current policy debate, as WTO President Mike Moore has done: "Sweeping generalisations are common from both the trade and the environmental community, arguing that trade is either good for the environment, full stop, or bad for the environment, full stop, while the real-world linkages are presumably a little bit of both, or a shade of grey" (qtd. in Economist 1999b). Although this study makes the case that trade hurts the environment, it is also recognized that trade is necessary and will be a continuing feature of the international economic system for the foreseeable future. That said, there are doubtless policies that balance the competing requirements of economic development and the global environment. As pointed out by Forrester (1971), they are probably nonintuitive and subtle, thus they effectively escape consideration in today's polarized policy environment. A debate structured by the analytic rigor of complex systems modeling would have a better chance of reconciling the disparate, polarized positions of economists and environmentalists and pinning down the analytically intermediate, "shades of grey," in a defensible fashion.

Although the ELP model is offered as a spare representation for use in exploring theory rather than a rich model for developing policy, it is still sufficiently detailed to critique at least one of the policies proposed in Seattle, that of further opening the markets of rich, northern nations so that developing southern nations can export more of their goods and become wealthy. Quite beyond the fact that the United States' trade deficit has grown from about $100 billion per year in 1992 (cf. figure B.6) to $1 billion per day less than a decade later (Norris 2001), the global environment is already severely degraded and probably cannot handle the whole world living at the Northern standard. The persistence of this policy prescription can be explained by the economist's belief that economic wealth leads to environmental health. The TC × GNP variable and ELP model demonstrate that the environmental health of developed nations comes about not because of their superior technology but because of their ability to push off the environmental consequences of their living standards through trade. Should southern nations develop to the northern standard, to which countries will they push off their environmental costs? The demonstrated empirical correlation between economic wealth and environmental health is predicated on asymmetric trade relationships. Therefore, to expect developing countries to grow their way to environmental health is not a viable scenario as there are no poorer countries with which they can trade asymmetrically.

In formulating policies that account for the health of both the world economy and the global environment, the key is to recognize that the complex international system will tend to undermine the corrective measures applied. Moreover, effective policy levers will be hard to locate and seem counterintuitive once they are found. Finally, there will be an ongoing tension between short-term and long-term benefits, especially in democracies. In formulating policy in such an environment, it helps to have a model on which proposed policies and possible scenarios can be simulated. Choucri (1981) provides such an example with her work on international petroleum markets. Economics teaches that price is a function of supply and demand. Choucri, in determining the price of oil, *disaggregates* international supply and demand so that the total world supply is broken up into Saudi Arabia, Kuwait, Venezuela, Nigeria, and other oil-producing states. World demand is also broken up into Western Europe, the United States, Japan, and other oil consuming states. Moreover, other oil market influences are included such as production and distribution details. The actual factors are less important than the fact that the international system can be disaggregated and connected in ways that provide greater fidelity to its actual features and insight into its actual behavior. Choucri (1981) provides greater insight into the international factors that drive the price and availability of oil; a similar modeling effort would provide insight into policies that balance the competing demands of economic development and the global environment. Ways to achieve this are explored in the next section.

6.4 Next Steps

In advocating the use of computer-based models to study complex social systems and formulate policies that apply to them, it is not intended that a technocracy replace the decision-making institutions presently in place. However, in a world of increasing globalization, it is clear that the complexity of the social systems being created is stretching the limits of present analytical and policy-making capabilities. The stridency on display in Seattle provides evidence for this conclusion. Each of the participants—the trade delegates from rich and poor nations, the labor activists, and the environmental protesters—held legitimate points of view from their respective perspectives. The problems they care about—grinding poverty, falling labor standards, and pervasive environmental degradation—are all recognized as germane and important. What is needed is a way to synthesize and reconcile these disparate and sometimes competing perspectives in a way that not only gives each its proper place and weight but also yields workable long-term solutions. The analytical techniques necessary for such a study are currently available and have been presented herein.

This focus of this study has been on trade and the environment. However, creating models of sufficient detail and complexity to formulate policy will require incorporating additional information and addressing additional topics. Doing so entails the exploration of other possible environmental relationships, and several are presented. First, there is the relationship between corruption and environmental degradation. In contrasting and comparing trade to the environment, and finding evidence against trade, one risks coming across as overly critical of economics. However, economists have long argued that the presence of subsides accompanied by the absence of enforceable property rights drives inefficient resource use and environmental degradation. For example, Brazil's deforestation problem ranks among the worst in the world and can be traced to a combination of poor economic policies, lawlessness, and insecure land tenure (Litvin 1998b). It might also be considered that since trade is associated with environmental degradation, and environmental degradation with corruption, might there be a link between trade and corruption? This question, while provocative, is left unexamined.

Another related topic is that between conflict and environmental degradation (Wils, Kamiya, and Choucri 1998; Homer-Dixon 1999). Can the decline of the natural environment affect the social environment to such an extent that it engenders violence? Stated another way, can changes in the natural environment lead to changes in the social? This study has focused on trade and the environment and has formulated its driving research questions and quantitative analyses to discover the extent to which the social environment has affected the natural. In developing empirical evidence to support the assertion that trade hurts the environment, it was necessary to focus on the expansion of the social environment at the expense of the natural. Feedback relationships have been used explicitly and extensively, so considering the extent to which the natural environment affects the social falls within the analytical scope of this study. This relationship manifests itself when the degradation of the natural environment causes such stress in the social environment that violence results. More empirical research on this topic in the form of statistical and complex system modeling is required.

The costs of globalization are not all environmental. Kristof (1999)[54] recounts the East Asian economic crisis of 1997. Economic shocks moved from country to country like a contagion touching Japan, Thailand, and Indonesia among others. What made this international transfer of trouble possible were the financial interconnections among recently deregulated Asian economies. Although the particulars of this crisis are essentially economic, the underlying analytic substance remains fundamentally the same—unintended consequences deriving from newly formed, complex social systems. The supporters of globalization, led by the United States and President Clinton's administration, recommended that the national economies of developing nations be opened so that investment capital from developed countries could flow in and build emerging markets. This in turn would increase the incomes of the poor in the developing world. What actually happened was a rapid influx of investment capital seeking high returns, the creation of an investment bubble when too much capital chased too few investments, and a rapid withdrawal of capital when the bubble burst. Some countries, especially Indonesia, were hard pressed to deal with the consequences of these economic shocks.

Globalization, like industrialization before it, is a disruptive social process that invariably creates winners and losers. But it is naive to expect that an increasingly integrated world will invariably provide more benefits than costs. Recall that one of the underappreciated aspects of complex social systems are tradeoffs between long-term and short-term benefits. Projects of political and economic integration tend to begin by tackling the easiest and most beneficial problems first.[55] Such easy and high return activities include establishing communications, travel, trade, and low tariffs. However, after the easy international connections have been established, and after people get a taste for integration and globalization, the more difficult problems begin to emerge such as migration, labor standards, and environmental degradation. This observation provides another interpretation of Seattle: The international system appears to be at the verge when the costs of globalization start to outweigh the benefits.

Addressing the costs of globalization will not be easy, but it is clear that international integration has brought with it unintended consequences of sufficient complexity to trouble traditional analytic methods. This being the case, it makes sense to apply techniques with enough analytic power to handle these complicated problems. Several methods have been used herein to study trade and the environment, and they can be extended to address these and other policy concerns. However, in doing so it is important to set realistic expectations. The simulation techniques proposed will not provide perfect prediction. They will be able to analyze more complex systems and incorporate more relevant theories and data than can be done now, which will aid in the understanding of globalization's true costs and foster the formulation of effective policies. The goal is not perfection. The goal, as in any scientific discipline, is to develop a methodology that refines the answers and draws close to an honest and comprehensive understanding. The systems study presented herein aids in this effort.

Appendix A
Complex Structures and Dynamics

Increases in computational power have allowed researchers to delve more deeply into the microfeatures and macrobehavior of complex systems generally and of natural systems specifically. This appendix begins by looking at the microfeatures of natural systems in the form of system structure or *fractals*. Next, the macrobehavior of complex, natural systems is examined in the form of system dynamics or *chaos*. Finally, the relationship between the social and natural environments is revisited in light of this analysis recognizing that variations of the techniques presented here are used in the primary analysis (cf. section 1.3 and chapter 5).

Mandelbrot (1991) proposes a theory of fractals that describes nature's basic geometry or structure. Figure A.1 shows perhaps the best known fractal, the Mandelbrot set (Schol 1995, frames #123 & #528). The Mandelbrot set is described mathematically by the equation $z(n+1) = z(n) * z(n) - z(0)$ where z is a complex number such that $z(0) = x + iy\ \forall (x, y)$ in the complex plane (where $i = \sqrt{-1}$). Two pictures are shown, and both demonstrate the characteristically multi-bubble, Mandelbrot shape. However, the difference in scale between the left and right frame is approximately the same as that between the Continental United States and a cell in one's finger. In fact, the diagram on the right was obtained by "diving" into the diagram on the left. These two pictures demonstrate one of the most basic characteristics of fractals specifically and of nature generally, self-similar structure. The larger-scale left diagram is actually composed of countless smaller-scale structures that can be investigated by mathematically diving into the picture. In doing so, it is impossible ever to get to "the bottom" because equally complex and sometimes similar structures emerge as the scale shrinks.

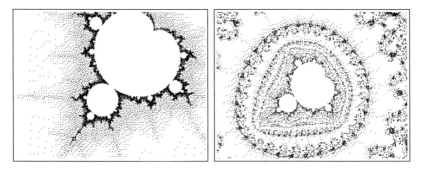

Figure A.1
Fractal Structure. Fractals exemplify *self-similar structure*. That is, large-scale fractals tend to look like small-scale fractals. Using the well-known Mandelbrot curve, even though the frame on the left is bigger than the frame on the right by many orders of magnitude, both frames evince the same characteristic, multi-bubble, Mandelbrot shape. Exploring fractal structures has only recently become possible with the advent of modern computers.

The discovery of fractals says something important about nature in that reality consists of myriad microfeatures that combine to form ever larger scale structures and objects. To think about this intuitively, consider how one tells the difference between a photograph and an animation. A photograph exhibits fine, small-scale surface details and consistent shading throughout the image, while an animation tends to be characterized by the absence of small-scale variation and lighting integrity. The computational difference between the two becomes evident when incorporating ever greater detail and integrity into a computer generated image—simple images can be rendered quite quickly, while complicated ones can take several hours even on fast computers. Because fractals are complex structures requiring many separate calculations to generate an image, their discovery became possible only in the era a powerful computers. In fact, the Mandelbrot set was discovered in 1979, well after powerful research computers became widely available (Reddy 1996, 91).

Computation has also proved important in revealing a near universal property of systemic macrobehavior, commonly called chaos. Although it might be expected that an argument will be made describing a close relationship between fractals and chaos, such is not the case. The strongest statements that can be made regarding fractals and chaos is that both are

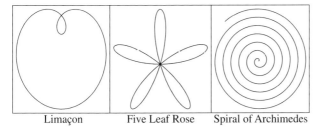

Limaçon — Five Leaf Rose — Spiral of Archimedes

Figure A.2
Algebraic Curves. These three algebraic curves represent the computational simplicity of the eighteenth century in that they can easily be drawn by hand. Continuing to draw the two curves on the left—the limaçon and five leaf rose—adds no new information. The spiral on the right, in contrast, continues to occupy space as it is drawn.

computationally expensive, and "many chaotic attractors exhibit fractal structure" (Bradley 1995, 759). Described by Prigogine and Stengers (1984) as nothing less than a new dialogue between man and nature, chaos denotes the complex, unpredictable, and seemingly random behavior that arises from deterministic physical systems. So ubiquitous is chaos that its study has been likened to the study of nonelephant animals (Stewart 1989, 81–84)—it is the nonchaotic systems are actually exotic and rare, not the chaotic. To understand the contribution of chaos, consider the curves of figure A.2.

The formulae for these curves are as follows (Leithold 1976, 566–567):

Limaçon: $\quad r = 2 - 3\sin\theta$

Five Leaf Rose: $\quad r = 4\sin 5\theta$

Spiral of Archimedes: $r = \theta$.

The limaçon and five leaf rose are types of algebraic curves, the drawing of which was popular in the early eighteenth-century due to the development of calculus and coordinate geometry (Rucker 1987, 132). These two curves, when drawn out dynamically on a computer, are initially interesting but quickly lose their appeal as the computer simply continues to draw over the same curve, endlessly. Phrased another way, the curves do not develop over time: once drawn, doing so again adds no new information. While this may be a drawback in the twentieth-century, it was a boon in the eighteenth because with no computers, the calculating and graphing of

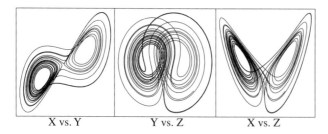

X vs. Y Y vs. Z X vs. Z

Figure A.3
Chaotic Curves: Lorenz Attractor. These three chaotic curves, when contrasted with the previous algebraic curves, are more complex as they evince a characteristic shape yet never repeat, rather like musical variations on a theme. Like fractal structures, the exploration of chaotic curves has only recently become possible with the advent of modern computers.

curves had to be done by hand. The third curve, the spiral of Archimedes, is exceptional in that it does not repeat, thus it alone offers a sense of progress over time and occupation of space. Spirals are in some sense less artificial than the other curves as they frequently occur in nature, such as snail shells, mountain goat horns, and so forth (Rucker 1987, 148–155).

Figure A.3 shows the chaotic curves developed by Edward Lorenz in the 1960's when modeling cloud dynamics (Gleick 1987, 11–31; Bradley 1995, 755). Visually, these curves combine the curves found in figure A.2 as chaotic curves simultaneously exhibit the constrained space of the limaçon and five leaf rose and the nonrepeatability of the spiral. Nonrepeatability is the first characteristic of chaos: no matter how long the curve is allowed to progress and develop, it will not repeat. Second, chaotic curves combine uniqueness and pattern in a natural way. While drawing a chaotic curve demonstrates a unique, characteristic pattern—the three curves within figure A.3 are all obviously different from one another—it also presents continuously new information representing progress or development. One might think about this in terms of people: each person is indeed a specific individual, but so too are they generally human. Third, the equations,

$$dX = -3(X - Y)$$
$$dY = 26.5X - Y - XZ$$
$$dZ = XY - Z,$$

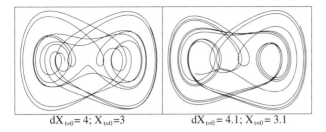

$dX_{t=0} = 4; X_{t=0} = 3$ $dX_{t=0} = 4.1; X_{t=0} = 3.1$

Figure A.4
Chaotic Curves Compared: Duffing Oscillator. These two curves, the left and right, are obviously different, but they are generated from the same set of equations. Their differences are due to separate starting points or *initial conditions*. That small initial differences eventually yield large consequences is a characteristic of chaos.

generate the Lorenz curves given the following initial conditions: $X_{t=0} = 0$, $Y_{t=0} = 1$, and $Z_{t=0} = 0$ (Thompson and Stewart 1986, 212–214). These three equations compose multiloop, nonlinear feedback systems of the type developed in chapter 5. Nonlinearity and feedback are key attributes of natural, chaotic systems (Bradley 1995, 757), just as they are in social systems (Forrester 1971). This means that the underlying mathematical structures leading to the dynamic behavior of both social and natural systems are the same (cf. Homer-Dixon 1993, 47–48).

Figure A.4 demonstrates another characteristic of chaos, sensitivity to initial conditions. The following equation describes the Duffing oscillator (Thompson and Stewart 1986, 3–5):

$$d^2 X + 0.05 dX + X^3 = 7.5 \cos\theta.$$

The two curves within figure A.4 are generated from the same equation; they are just started with slightly different initial conditions. The nonlinearity and feedback of the chaotic system amplify small initial perturbations so that physical systems starting at similar, though not equal, initial conditions end up very far from one another after a time (i.e., over several oscillator cycles or periods). Through chaos, deterministic physical systems can yield probabilistic dynamics. Part of the reason for this comes from background noise, which causes chaotic systems to jump between trajectories. Noise blends seamlessly into a chaotic system, rather like an additional shuffle would affect an already shuffled deck of cards

(Bradley 1995, 770). Were the deck rigidly ordered, then shuffling would produce much more drastic effects. For deterministic digital systems that resist noise—like a computer—starting from the same initial value always generates the same curve.

Prigogine and Stengers (1984) move a good deal beyond empirical and mathematical descriptions of chaos to address its philosophical implications. They maintain that chaos rejects the reductionism of traditional science and embraces a more holistic and systemic conception of nature. Doing so opens up a "new dialogue" between humanity and nature. This study reaches the same conclusion if by dialogue one means discovering and defining interconnections between the social and natural environments. This study has argued that both environments, the social and natural, can be represented through multiloop, nonlinear feedback models. Using the newfound power of the computer, such models make explicit the relationships between a system's microfeatures and its macrobehavior.

Appendix B
Conditional Plot Analysis

This appendix provides a third level of analysis to the time-series discussion of chapter 3. Two levels of analysis were examined in section 3.2—the (1) global and (2) regional as divided into the developed North and developing South. Here, the regional analysis is decomposed still further as the North and South are split out by *profile* per the table 3.1 definitions. The key distinction is that between relative global shares of population and GNP: groups 1, 2, and 3 have a greater global share of population than GNP and represent the developing South; groups 4, 5, and 6 have more GNP than population and represent the developed North. Within these two groupings, the greater the number, the more urbanized it is. Conversely, the lower the number, the more area. So profile 6 countries are more urbanized than profile 4 countries that have more area, and the same relationship holds for urban profile 3 countries and more landed profile 1 countries. The graphs are called *conditional plots*, with each of the six panels representing a profile—from profile 1 in the lower left to profile 6 in the upper right—as shown in figure B.1.

The data in these plots are not summed as they were previously, which makes the data more cross-national and comparable. Each point represents a country's value for a single year. The data are only ordered through the conditioning by profile group.

Figure B.2 presents the GNP data of figure 3.6 in a profile-based format using a conditional plot. Looking to the figure's content, only a few countries dominate: the United States in profile group 5 followed by Japan, Germany, France, and Great Britain in group 6 and China and India in group 2. While other countries indeed carry on significant economic

178   Appendix B

Profile 5	Profile 6
Profile 3	Profile 4
Profile 1	Profile 2

Figure B.1
Profile Pattern of Conditional Plots. The conditional plots all use this profile distribution pattern.

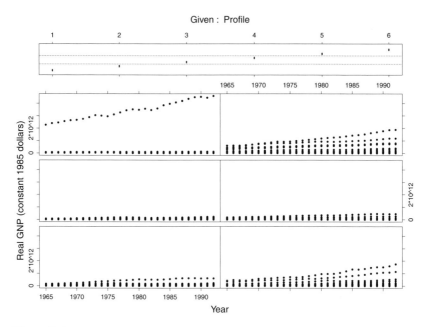

Figure B.2
GNP by Profile. Standout GNP countries include the United States in profile 5, Japan, Germany, France, and Great Britain in profile 6, and China and India in profile 2.

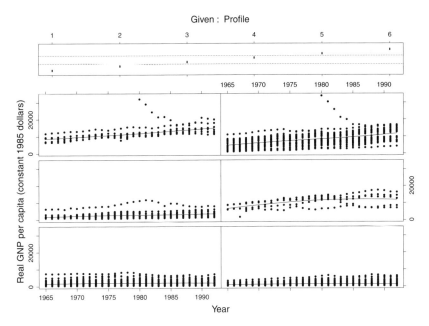

Figure B.3
GNP per Capita by Profile. Standout GNP per capita countries include Qatar, Luxembourg, Hong Kong, Switzerland, and Japan in profile 6, United Arab Emirates and the United States in profile 5, Canada and Australia in profile 4, Trinidad and Tobago in profile 3, Mexico in profile 2, and Venezuela in profile 1.

activity, their individual GNPs do not stand out relative to the larger economies mentioned previously. This applies especially to the countries in profile groups 1, 3, and 4.

Figure B.3 presents the GNP per capita data in a conditional plot, just as figure B.2 showed GNP data. Once again, the data are presented in a straightforward manner—sums or means are not taken, the data are only organized by profile group. However, a line denoting aggregate average has been included in the conditioning panels so that trends can be more easily determined (StatSci 1995, 9–10). Standout countries, including the United Arab Emirates (group 5) and Qatar (group 6), experienced radical increases in GNP per capita after 1980 due to the OPEC oil embargo. Other high GNP per capita countries include Luxembourg, Hong Kong, Switzerland, and Japan (group 6), the United States (group 5), Canada and Australia (group 4), Trinidad and Tobago (group 3), Mexico

180 Appendix B

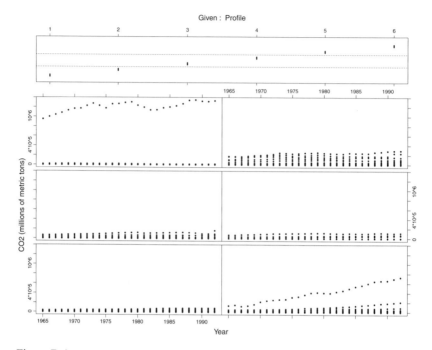

Figure B.4
CO_2 **by Profile.** Standout total CO_2 contributors include the United States in profile 5 and China in profile 2.

(group 2), and Venezuela (group 1). Comparing the profile groups themselves reveals that groups 4–6 all exhibit comparably high and growing living standards, while the countries within profile groups 1–3 all exhibit comparatively low standards. Finally, the large-scale national differences in absolute GNP (cf. figure B.2) become much smaller when compared by GNP per capita. This is especially true for the profile group 4 countries like Canada and Australia that have comparatively small absolute GNP's, but their GNP per capita values are quite comparable to those of profile groups 5 and 6.

The total CO_2 country contributions by profile, presented in figure B.4, show that the United States (profile group 5), China and India (profile group 2) dominate international CO_2 contributions. The former Eastern Bloc, including the Soviet Union, now Russia, is also a major contributor of CO_2, but it is not shown because these countries were not included in

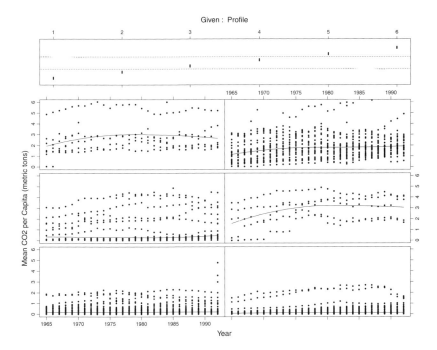

Figure B.5
CO₂ per Capita by Profile. The developed North, profiles 4–6, demonstrates much higher CO_2 per capita values (as denoted by the averaging line) than the developing South, profiles 1–3.

the North and Choucri profile groupings. Profile group 6 is dominated by Japan and Germany, but they don't set themselves apart from the rest of the profile group 6 countries, although group 6 demonstrates much higher effluent rates on average than groups 1–5, save the United States and China.

Figure B.5 shows that a clear split exists between North and South, as with GNP per capita. The northern profiles, 4–6, all show much higher though leveling values of CO_2 per capita, while the southern profiles, 1–3, all show much lower levels.

Figure B.6 reveals the United States to be the world's import leader, with the industrialized nations of profile group 6—Germany, France, Japan, and Great Britain following close behind. Other notable importers are Canada in group 4 and China and Mexico in group 2. The export portion of figure B.6 shows exports broken out by profile group, but the margin

182   *Appendix B*

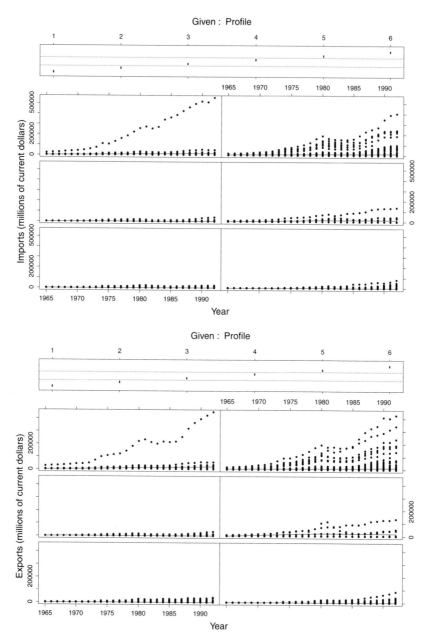

Figure B.6
Imports and Exports by Profile. For most countries, imports equal exports. The United States, which dominates profile 5, however imports far more than it exports.

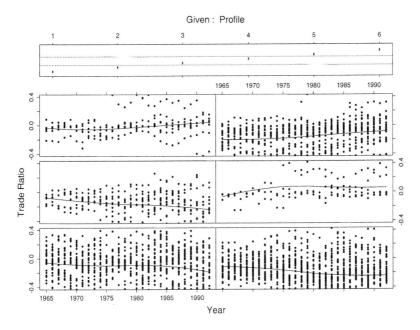

Figure B.7
Trade Ratio by Profile. The developed North, profiles 4–6, demonstrates growing exports as denoted by their positive trade ratio lines. The developing South, profiles 1–3, demonstrates growing imports as denoted by their increasingly negative trade ratio lines.

by which the United States led the group 6 countries has decreased. The United States exported about as much as Germany did in 1992, hence America's $105 billion 1992 trade deficit. In group 6, Germany is followed by Japan, France, and Great Britain. Other notable exporters are Canada in group 4 although it was briefly eclipsed by Saudi Arabia in 1980 and 1981; China and Malaysia stand out in group 2.

Comparing imports and exports is not particularly revealing as the two graphs are quite similar, imports tend to equal exports. Figure B.7 graphs the trade ratio, $(exports - imports)/(exports + imports)$, to help determine the terms of trade for each profile. More exports than imports yields a positive number between 0 and 1, while more imports than exports generates a negative value. The results are not too surprising considering the previous analysis: the trade ratios increase for the northern profile groups indicating a general increase in exports, while the

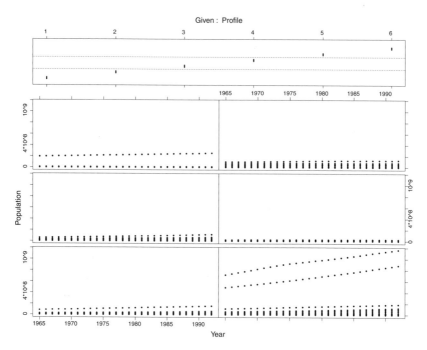

Figure B.8
Population by Profile. Standout total population countries include China and India in profile 2, and the United States in profile 5.

trade ratios for the South tend to fall off indicating a general increase in imports.

Figure B.8 shows the profile groups for population. Immediately obvious are the two dominant countries in profile group 2, China and India, that have large and growing populations. In fact, China sets the graph's scale with a population of over one billion. The only other country that sets itself apart is the United States in profile group 5 with a slowly growing population of over 250 million. Beyond that, it is difficult to glean individual country information from figure B.8, save the population preponderance of China and India.

Figure B.9 shows the distribution of population growth rates among the six profile groups. Profile groups 5 and 6 demonstrate growth rates of 1 percent; profile group 4, 2 percent; profile group 3, 2 to 1 percent; and profile groups 1 and 2, 2 percent. What is of interest here is

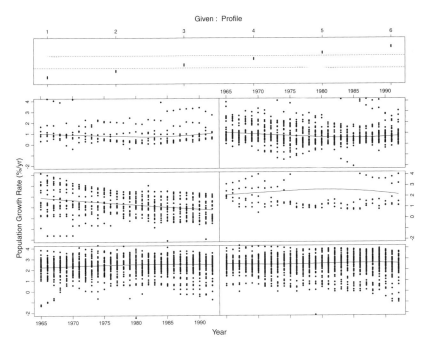

Figure B.9
Population Growth by Profile. The developing South, profiles 1–3, demonstrate higher population growth rates than the developed North, profiles 4–6. Note that the low area countries in profiles 3 and 6 show lower growth rates than those with more land (e.g., profiles 1 and 4).

not that the South demonstrates much higher population growth rates, which was expected, but the impact of resources or land area on the rates of growth. The two most resource constrained profile groups, 3 and 6, both show decreasing growth rates, especially profile group 3. Meanwhile, those profile groups with the greatest relative share of resources, 1 and 4, both have population growth rates of over 2 percent. This subtle influence provides an example of the natural environment affecting the social.

Figure B.10 is presented to help ascertain general trends in deforestation rates. It reveals a clear pattern for deforestation rates among the six profiles. The South, profile groups 1–3, exhibits large deforestation rates with considerable numbers of data points in the negative range. The North, profile groups 4–6, shows comparatively few negative values and

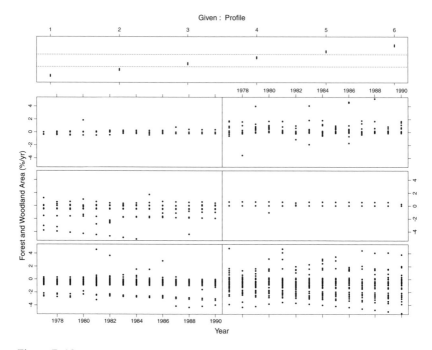

Figure B.10
Forest and Woodland Change by Profile. The developing South (profiles 1–3) show more deforestation, as denoted by negative forest change rates, than the developed North (profiles 4–6).

many positive ones. This leads to the question, are high domestic population growth rates the cause of deforestation, or is the cause the South's position within the international system? Both domestic and international explanations are important and plausible and are tested directly in section 4.4.

Figure B.11 shows the distribution of agricultural land by profile group. As noted previously, agricultural land appears to be increasing in the South, as exemplified by profile groups 1–3. Group 3, the most resource constrained of the southern profiles, shows a small increase, but the countries of profile group 2 appear to be taking advantage of their resources by dramatically increasing their agricultural lands. Turning to the northern profile groups 4–6, it appears that groups 4 and 5 remain essentially static over this period. Group 6, the most resource

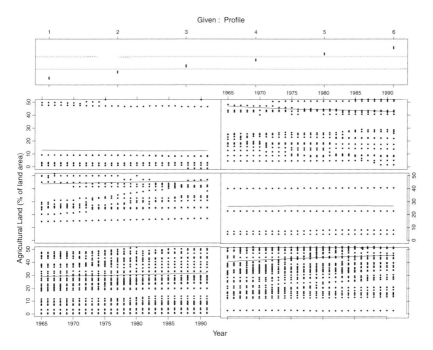

Figure B.11
Agricultural Land by Profile. Countries with little area, profiles 3 and 6, have higher percentages of agricultural land than those with more.

constrained of the northern profiles, shows the most dramatic declines in agricultural lands. Thus, as one would expect, agricultural expansion takes place in the profiles where there is land and population available for expansion, while decline takes place in the most resource constrained profiles groups.

Appendix C
Latitude and Longitude Analysis

This appendix continues the "four major variables as a function of absolute latitude" discussion presented in section 4.2. Recall that table 4.1 shows two variables, GNP per capita and CO_2, that increase as one moves *away* from the equator towards the North or South poles, and two other variables, population growth and deforestation, that increase as one moves *toward* the equator and away from the poles.

Tests were also run for signed latitudes, latitudes that increase as one moves from the South Pole ($-90°$) to the Equator ($0°$), and finally to the North Pole ($90°$), as well as for the four major variables yielding the following t-value/probability pairs (less than 5% is statistically significant): GNP per capita, 5.19 (0.0000); CO_2 per capita, 5.22 (0.0000); population growth, -5.22 (0.0000); deforestation, -2.05 (0.0404). Once again, a northern orientation for GNP per capita and CO_2 per capita and a southern orientation for population growth and deforestation is found. Note that the deforestation statistic (UNFAO 1993; UNFAO 1995; World Bank 1995) is a ten-year average taken between 1980 and 1990, so ten-year averages were calculated for the other variables as well. Initially the GNP per capita, CO_2 per capita, and population growth were calculated from the year 1990 generating t-value/probability pairs of 8.325 (0.0000), 7.211 (0.0000) and -2.931 (0.0034), respectively. Deforestation, of course, remains the same. The lower t-value for population growth is probably due to the 1990 migration outliers that get averaged away, for example, Bulgaria at -4 percent; Qatar $+14$ percent; Rwanda $+62$ percent.

Signed longitude tests, starting from $-180°$ in the Pacific and moving east over the Americas, hitting $0°$ in Western Europe, moving east over Asia, and then finishing at $180°$ in the Pacific again, were run for the sake of thoroughness and generated the following t-value/probability pairs: GNP per capita, 0.731 (0.466); CO_2 per capita, 0.851 (.397); population growth, .323 (.748); and deforestation, -1.89 (.0615). The deforestation result is somewhat surprising as it is close to being significant at the 5 percent level. Therefore, deforestation decreases as one moves east (or increases as one moves west). This indicates that deforestation occurs primarily in West Africa and the Western hemisphere. The same test is run for deforestation using the absolute value of longitude, longitude increasing as one moves away from Western Europe and toward the Pacific, either moving west over the Americas or east over Asia, yielding the following results: $\hat{\beta}_{|long|}$, 0.0159; σ, 0.0036; t, 4.47; pr, 0.0000. So, as one moves away from the Greenwich meridian (0° longitude) and toward the Pacific Ocean, either east or west, deforestation increases. This is probably because Southeast Asian deforestation is now counted positively along with deforestation in the Americas.

The high deforestation African states are clustered around 0° longitude, but they are probably outweighed and counterbalanced by low European deforestation. Note also that this parameter estimate, at a little more than half the parameter value of the absolute latitude test, compares favorably when one considers that there are twice as many degrees of longitude than latitude.

It is interesting to consider that this longitudinal effect may be an artifact of the world economy's spread from London in the nineteenth century (Braudel 1979), since 0° longitude is centered on the old Royal Greenwich Observatory site in Greenwich and Great Britain is a well known importer of timber (see Akira 1989). A quick test was run on deforestation as a makeshift function of distance from London which generated a t-value of 4.29 (0.0000), which is significant, but the $\hat{\beta}$ on which it is based, 0.0003, is not quite as large as those generated by the absolute longitude or latitude tests, indicating the need for additional analytical inquiry beyond the bounds of this study.

Appendix D
TC × GNP Tests

This appendix contains three tests of table 4.7. First, diagnostic plots of the model show limited explained variance with a low R^2, which is to be expected from a large, international dataset. More important is model fit, which is high as measured by the F-statistic (74.1). Second, a Panel Corrected Standard Errors (PCSE) analysis is performed, which explicitly accounts for temporal autocorrelation and heteroskedasticity. This test further supports the table 4.7 results. Third, a Moran's I cross-sectional analysis is performed that tests for spatial autocorrelation, the spatial analogue of temporal autocorrelation. These results show that spatial autocorrelation is not a problem for this model.

D.1 Diagnostic Plots

Figure D.1 contains a series of standard S-PLUS™ diagnostic plots that evaluate the explained variance of the model, Forest Change (delta Forests and Woodlands or dFW) as a function of Trade Connected GNP (TC × GNP), Local GNP Per Capita (LGNPC), and Population growth (delta Population or dPOP). Overall, figure D.1 reveals limited explained variance or low R^2. The top left plot shows residuals mapped against the fitted values generated by the model. Three outliers are labeled: from top to bottom they are Sri Lanka 1986 (#1030), Paraguay 1986 (#1437), and Reunion 1981 (#1447). All three are forest change outliers with values of 4.01 percent/year, −3.90 percent/year, and −3.98 percent/year respectively (mean forest change is −0.12 percent/year). Looking to the scale of the upper left plot's X and Y axes, the fitted values range between −0.4 and 0.1, while the residuals range between −4 and 4, indicating limited

192 *Appendix D*

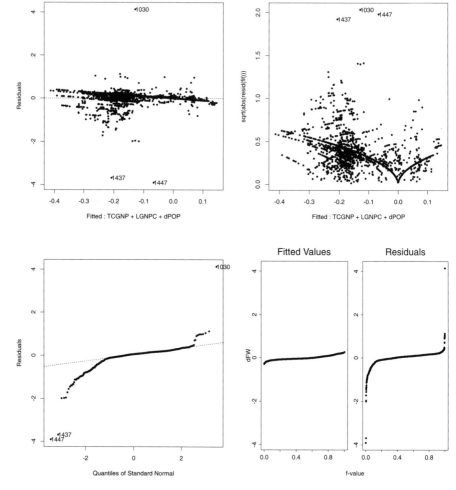

Figure D.1
Forest Change Model Diagnostic Plots. The diagnostic plots reveal limited explained variance in the table 4.7 model.

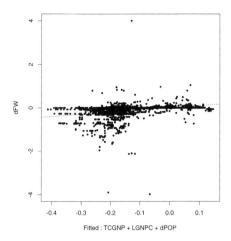

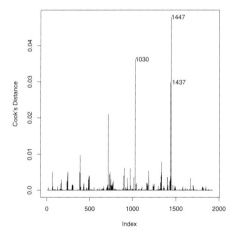

explained variance for the model. So too, the residuals continue to exhibit pattern featuring a cluster to the left that extends downward, another cluster to the right that extends upwards, and a more linear cluster of points that descend shallowly from the upper left to the lower right.

Looking to the upper center plot, the square root of the residual's absolute values are shown against the fitted values. This plot serves to further identify pattern within the residuals, and the previously identified patterns appear here even more starkly. The three outliers are now together and the largely negative left cluster has been flipped upward due to the absolute value function. The somewhat diminished cluster to the right also appears more starkly in this projection. This mismatch in residual cluster size between left and right indicates that there are many more cases of negative forest change than positive. This observation is further supported by the range of the fitted values, from -0.4 to 0.1. The curved lines that meet at the origin $(0,0)$ derive from the linear pattern of points in the previous plot. These appear curved due to the square root function. Note also that the vertical axis ranges between 0 and 2 instead of 4 due to the square root function. The curved line represents points of zero forest change, and their influence is significant; 35.6 percent of the forest change values in this model are zero. This occurs for two reasons: first, forest change may actually be zero, in which case the measurement needs to be included in the model; second, the same forest area value may be reported year after year with the variance between years going underreported. It might be useful to exclude the second category of zeros, but there is no way to discriminate between the categories and so all values are included.

As an experiment, the model was run with all forest change zeros excluded. As expected, the range of fitted values increased from -0.4 and 0.1 to -1.0 and 0.1; the R^2 increased from 8.5 percent to 10.6 percent. The fitted values' spread remained smaller than the residuals' because most of the observations remained clustered at low forest change values. Thus, the zero values are only one aspect of forest change underreporting.

The upper right plot shows the dependent variable, dFW, plotted against the fitted values with the regression line included, revealing a plot that looks similar to the first residual plot except that it has been rotated several degrees to the left. This indicates that the residuals are correlated with the dependent variable. Testing returns a correlation of 0.959, which corresponds to an R² of 0.085 or 8.5 percent. This low explained variance can be interpreted in two ways. First, the large number of forest change zeros makes the slope of the regression line much shallower than it would be otherwise. A steeper slope would explain more of the variance, the fitted values would exhibit a wider range, and more of the pattern would be absorbed from the residuals. Stated simply, better forest change data would help. More generally, when working with time-series, cross-sectional data from something as complex as the international system, it would be unreasonable to expect that a few variables could explain large amounts of variance. The difficulty is twofold as there exist both data and theoretical limitations. However, the reported F-statistic of table 4.7, 74.14, reveals strong model fit, and since the data limitation exists in the dependent variable, the three independent variables all face the same explanatory challenge. Therefore, while the scale of the explanatory parameter estimates may be reduced, relative comparisons among the estimates remain justified and valid.

The lower left and center panels show the normal quantile plot of residuals and the residual-fit spread plot. The quantile plot ideally would show the residuals clustered closely against the dotted line revealing a normal distribution. Instead, the long tails and shallow slope demonstrate this is not the case. The residual-fit plot ideally would show greater spread for the fitted values than the residuals. Once again, this is not the case. Finally, the lower right plot contains Cook's distance information, which identifies the most influential observations. The identified three are Sri Lanka 1986, Paraguay 1986, and Reunion 1981, all of which are forest change outliers. Removing these three observations improves the t-ratio somewhat for the three parameter estimates, but beyond that the results remain essentially unchanged.

To conclude, figure D.1 reveals low explained variance, which is to be expected from a system as vast and complex as the international system. However, the F-statistic demonstrates strong model fit, and since the data limitation exists in the dependent variable, forest change, the three independent variables, trade connected GNP, local GNP per capita, and population growth, all face the same explanatory challenge. While the scale of the parameter estimates may be reduced, comparisons among the three explanatory variables remain justified and valid.

D.2 Temporal Autocorrelation

It is generally known that forest change data tend to be highly autocorrelated over time, and so the multivariate model was tested with a temporally lagged dependent variable (i.e., last year's value for forest change was included as a fourth independent variable). The lagged variable generated a parameter of value 0.91 with the other three parameter values rendered statistically insignificant. This means that the previous year's forest change value does a better job at predicting the next year's value than do the combination of trade connected GNP, GNP per capita, and population growth. To account for the time-based difficulties introduced by temporal autocorrelation and heteroskedasticity, Panel Corrected Standard Errors or PCSEs (Beck and Katz 1995) are computed for the model. This test delivers results similar to those of table 4.7, which supports the statistical analysis of chapter 4. Because PCSEs do not permit missing data, all countries with missing observations in the fifteen-year panel timeframe between 1976 and 1991 are omitted, reducing the number of observations and changing the parameter estimates. The focus of PCSE however is on the standard errors that are recalculated to take into account the variance lost from pooling. In other words, PCSEs account for time within the model.

Table D.1 shows the results of table 4.7 persist. First, the number of relevant data points was reduced from 1929 to 1665. In culling the data set to obtain testable data, there exists the danger of throwing out the developing countries and focusing on the developed ones that tend to have better data. This is not the case however as GNP per capita decreases from 4534 to 4213, and population growth increases from 1.97 to 2.12—both of which indicate a slight shift towards developing rather than developed nations.

The parameter estimates tell the same story with the estimate for TC × GNP increasing from -0.121 to -0.139, GNP per capita decreasing slightly from 13.9×10^{-6} to 13.3×10^{-6}, and population growth increasing from -0.0391 to -0.0531. These changes are reflected in the mean multiples with TC × GNP and population growth both negative and close to the same magnitude of mean forest change (95% and 78% respectively), while GNP per capita remains less than 40 percent of mean forest change.

Table D.1
Panel Corrected Standard Errors (PCSE) Test

Coefficients	$\hat{\beta}$	σ	t	pr(>|t|)
(Intercept)	0.0493	0.0249	1.98	0.0479
TC × GNP	−0.139	0.0151	−9.19	0.0000
GNP per capita	13.3μ	1.28μ	10.4	0.0000
Population growth	−0.0531	0.00747	−7.11	0.0000

pr(>|t|) denotes probability of null hypothesis; $\mu = 10^{-6}$.
F-stat.: 80.49 on 3 and 1661 degrees of freedom; the prob. is 0.

mean(TC × GNP) = 0.980		T$(1985)/year
mean(GNP per capita) = 4213		$(1985)/person-year
mean(Population growth) = 2.12		%/year
mean(Forest change) = −0.143	(1.00)	%/year
$\hat{\beta}_1$ × mean(TC × GNP) = −0.136	(0.95)	%/year
$\hat{\beta}_2$ × mean(GNP per capita) = 0.0561	(−0.39)	%/year
$\hat{\beta}_3$ × mean(Population growth) = −0.112	(0.78)	%/year

Values in parentheses denote ratio to mean Forest change; $T = 10^{12}$.
Note: PCSEs correct for autocorrelation and heteroskedasticity in table 4.7, and TC × GNP remains a key explanatory variable of forest change.

The PCSE test confirms the table 4.7 results in the presence of autocorrelation and heteroskedasticity. However, using PCSEs implies a more subtle point as well. More than simply treating time dependence as a nuisance to be removed, PCSEs assert that time dependence is an important aspect of time-series, cross-sectional studies; the PCSE technique acknowledges rather than excises temporal dynamics as doing otherwise can mislead the researcher and undermine the analysis (Beck and Katz 1995, 27).

D.3 Spatial Autocorrelation

When performing tests for temporal autocorrelation in a time-series, cross-national dataset, one must also consider the possibility of spatial autocorrelation. If temporal autocorrelation allows one to predict a value by its preceding value, then spatial autocorrelation allows one to predict a value by its neighboring value. This topic has already been addressed descriptively through use of the G_i^* statistic (Ord and Getis 1995) in

Table D.2
Cross-Sectional Tests

Year	TC × GNP	GNP per capita	Population growth	Moran's I
1977	−0.246 (0.00)	19.0μ (0.05)	0.00119 (0.97)	(0.17)
1978	−0.231 (0.00)	23.9μ (0.01)	0.00228 (0.94)	(0.17)
1979	−0.240 (0.00)	22.2μ (0.02)	−0.00502 (0.87)	(0.22)
1980	−0.200 (0.00)	18.0μ (0.00)	−0.0527 (0.00)	(0.10)
1981	−0.204 (0.00)	17.0μ (0.01)	−0.0515 (0.03)	(0.04)
1982	−0.190 (0.00)	16.6μ (0.01)	−0.0435 (0.07)	(0.09)
1983	−0.155 (0.01)	16.4μ (0.02)	−0.0434 (0.08)	(0.11)
1984	−0.102 (0.04)	12.7μ (0.05)	−0.0418 (0.08)	(0.86)
1985	−0.123 (0.01)	11.3μ (0.08)	−0.0469 (0.06)	(0.34)
1986	−0.176 (0.08)	7.83μ (0.59)	−0.101 (0.07)	(0.83)
1987	−0.101 (0.05)	13.1μ (0.08)	−0.0503 (0.09)	(0.75)
1988	−0.112 (0.02)	12.4μ (0.09)	−0.0501 (0.09)	(0.48)
1989	−0.120 (0.02)	7.60μ (0.32)	−0.0650 (0.05)	(0.31)
1990	−0.109 (0.02)	12.1μ (0.06)	−0.0557 (0.04)	(0.39)
1991	−0.114 (0.01)	12.2μ (0.03)	−0.0473 (0.04)	(0.46)

Values in parentheses denote probability of null hypothesis; $\mu = 10^{-6}$.
Note: A mean Moran's I probability of 0.35 denotes some but statistically insignificant spatial autocorrelation.

figures 4.1, 4.2, 4.3, and 4.4, and some spatial autocorrelation was visually evident within each variable. Here, Moran's I probabilities (Cliff and Ord 1981, 42–51) are used to evaluate spatial autocorrelation in the table 4.7 multivariate model. Moran's I probabilities denote the likelihood of the null hypothesis, that there is no spatial autocorrelation in the model. Probabilities less than 5 percent denote statistically significant spatial autocorrelation. This section finds that spatial autocorrelation is not a problem for the model.

Table D.2 shows the initial results of the Moran's I tests. The model is broken into slices by year and Moran's I probabilities are calculated for each slice. Before turning to spatial autocorrelation for the final analysis, it must be noted that the yearly slices yield interesting results. First, the slices all show good fit according to the the F-statistic with all but one year below the 1 percent probability level, the exception being 1986 at 3.5 percent. Looking to the parameter estimates, TC × GNP is significant at the 5 percent level fourteen out of fifteen years, GNP per capita nine times, and population growth five times. The Moran's I probability of spatial autocorrelation generates semi-low values indicating the presence of some spatial autocorrelation interspersed with high values indicating its absence. Only one of the fifteen values is statistically significant, and at 4 percent barely so, indicating that spatial autocorrelation is not a significant problem in the model. The mean Moran's I probability for the fifteen years tested is 0.35.

Including a temporally lagged variable in a model accounts for temporal autocorrelation, and including a spatially lagged variable accounts for spatial autocorrelation. Just as a temporally lagged variable denotes the value that came before it in time, a spatially lagged variable is determined by the values that surround it in space. For example, the spatially lagged deforestation value for the United States would be determined by the deforestation rates of Canada and Mexico. Operationally, the spatially lagged variable is calculated by cross-multiplying a contiguity matrix (concept introduced in section 4.2) with the forest change values for that year. The results from including a spatially lagged variable are shown in table D.3. The slices continue to show good fit using the F-statistic with all but one year below the 1 percent probability level, the exception once again being 1986 at 6.2 percent. Given these exceptions, the fit seems to be better across the board by a couple of orders of magnitude with respect to percentage or four full F-statistic points (mean F-statistic increases from 6.4 for table D.2 to 10.7 for table D.3). The Moran's I probabilities also tend to be higher across the board demonstrating an increased accounting for spatial autocorrelation in the model through the lagged variable. The mean Moran's I probability for table D.3 is 0.54, indicating less latent spatial autocorrelation than the 0.35 from table D.2.

Spatial autocorrelation was also tested using the trade matrices rather than the contiguity matrix. Whereas the contiguity matrix-based analysis generated a fifteen-year mean Moran's I probability of 0.35 indicating the presence of some spatial autocorrelation, the trade base analysis yielded a range of probabilities between 0.75 and 0.99 indicating its almost total absence.

Looking to the table D.3 parameter estimates, lagged forest change is the most stable variable with 11 of the 15 years significant at the 5 percent

Table D.3
Lagged Spatial Weight Tests

Year	Forest change (lag)	TC × GNP	GNP per capita	Population growth	Moran's I
1977	0.767 (0.00)	−0.0780 (0.26)	12.1μ (0.16)	0.0206 (0.47)	(0.35)
1978	0.800 (0.00)	−0.0653 (0.30)	15.5μ (0.07)	0.0255 (0.35)	(0.39)
1979	0.710 (0.00)	−0.0824 (0.20)	15.4μ (0.06)	0.0182 (0.52)	(0.36)
1980	0.508 (0.00)	−0.120 (0.02)	11.6μ (0.02)	−0.0388 (0.02)	(0.47)
1981	0.966 (0.00)	0.00399 (0.94)	3.14μ (0.53)	−0.0286 (0.11)	(0.20)
1982	0.858 (0.00)	−0.0282 (0.62)	4.00μ (0.48)	−0.0285 (0.15)	(0.19)
1983	0.629 (0.00)	−0.0674 (0.23)	7.13μ (0.27)	−0.0366 (0.10)	(0.35)
1984	−0.0832 (0.59)	−0.112 (0.04)	13.7μ (0.04)	−0.0428 (0.08)	(0.60)
1985	0.396 (0.01)	−0.0684 (0.17)	6.48μ (0.32)	−0.0406 (0.09)	(0.54)
1986	0.174 (0.54)	−0.150 (0.17)	5.21μ (0.73)	−0.0971 (0.09)	(0.91)
1987	0.115 (0.55)	−0.0855 (0.14)	11.6μ (0.14)	−0.0488 (0.10)	(0.98)
1988	0.326 (0.07)	−0.0681 (0.21)	8.15μ (0.28)	−0.0464 (0.11)	(0.70)
1989	0.361 (0.03)	−0.0750 (0.16)	4.38μ (0.57)	−0.0563 (0.08)	(0.65)
1990	0.339 (0.04)	−0.0690 (0.16)	8.50μ (0.20)	−0.0475 (0.08)	(0.70)
1991	0.329 (0.05)	−0.0813 (0.06)	8.83μ (0.13)	−0.0415 (0.07)	(0.68)

Values in parentheses denote probability of null hypothesis; $\mu = 10^{-6}$.
Note: A mean Moran's I probability of 0.54 indicates the spatially lagged variable, Forest Change (lag), accounts for some spatial autocorrelation.

level and a mean probability of 13 percent. TC × GNP and GNP per capita both are significant 2 of 15 years at the 5 percent level. TC × GNP changes sign in 1981, but with a null probability of 94 percent the estimate is functionally zero. The mean probability of TC × GNP is 25 percent, and that of GNP per capita is 27 percent. Population growth, in contrast, has only one estimate significant at the 5 percent level (1980) and features three years with incorrect sign. However, its mean probability is 16 percent.

Appendix E
Dynamic Model Equations

It is important that the details of quantitative models be provided so (1) their analytic assumptions are explicit and supported, and (2) their results can be replicated if desired. In the spirit of explicitness and replication, this appendix presents the differential equations of the two system dynamics models presented in chapter 5, the overshoot and collapse model and the environmental lateral pressure model. The equations for each variable and relationship are provided as are short verbal descriptions. Sufficient information is given that both models could be replicated on another simulation system. Nonlinear, table-based relationships are provided in this appendix, although their numerically based descriptions are hard to visualize. Therefore, more developed and graphically explicit descriptions are also provided in appendix F.

E.1 Overshoot and Collapse Model

For full details of this model, see HPS (1990, chap. 9).

$Food(t) = Food(t - dt) + (-consumption) * dt$, INIT $Food = 100$
DOCUMENT: The initial value for Food is 100.

$consumption = Population * consumption\ per\ day$
DOCUMENT: Consumption is determined by multiplying Population by consumption per day.

$Population(t) = Population(t - dt) + (births - deaths) * dt$, INIT $Population = 2$
DOCUMENT: The initial value for Population is 2.

births = Population ∗ birth fraction
DOCUMENT: Births is determined by multiplying Population by *birth fraction*. Since *birth fraction* is constant, the larger the Population, the larger the births flow. This is a compounding flow.

deaths = Population ∗ death fraction
DOCUMENT: Deaths is determined by multiplying Population by *death fraction*. This is a draining flow.

birth fraction = 0.2
DOCUMENT: The *birth fraction* variable is set at a constant 20 percent per year.

food per capita = Food/Population
DOCUMENT: *food per capita* is determined by dividing Food by Population.

consumption per day = GRAPH(food per capita) (0.00, 0.00), (0.2, 0.36), (0.4, 0.57), (0.6, 0.725), (0.8, 0.775), (1.00, 0.795), (1.20, 0.8), (1.40, 0.8), (1.60, 0.8), (1.80, 0.8), (2.00, 0.8)
DOCUMENT: *consumption per day* is defined as a graphical function in which *consumption per day* is a function of *food per capita*. *consumption per day* tends to increase as *food per capita* increases.

death fraction = GRAPH(food per capita) (0.00, 0.5), (1.00, 0.3), (2.00, 0.075), (3.00, 0.0275), (4.00, 0.015), (5.00, 0.015), (6.00, 0.015), (7.00, 0.015), (8.00, 0.015), (9.00, 0.015), (10.0, 0.015)
DOCUMENT: *death fraction* is a graphical function of *food per capita*. An increase in *food per capita* causes *death fraction* to decrease.

E.2 Environmental Lateral Pressure Model

The environmental lateral pressure equations make use of two sets of identical equations, one for the developed North and another set for the developing South. To demonstrate their similarity, both North and South equations are presented together and share verbal descriptions. They are differentiated by a suffix at the end of the variable name. For example, *investment* is replicated and renamed *investment.N* for North and *investment.S* for South. Note that the variable's units are given in parentheses after the variable's name in the Document line. Empty parentheses denote dimensionless values.

$GNP.N(t) = GNP.N(t - dt) + (investment.N - depreciation.N) * dt$,
INIT $GNP.N = 2500 * Population.N$
$GNP.S(t) = GNP.S(t - dt) + (investment.S - depreciation.S) * dt$, INIT $GNP.S = GNP.N$
DOCUMENT: Gross National Product ($/year)—Initial GNP, a proxy measure of technology. This value represents society's infrastructure (i.e., its ability to provide goods and services over one year).

$investment.N = (invest\ rate.N + tech\ trade.N) * GNP.N$
$investment.S = (invest\ rate.S + tech\ trade.S) * GNP.S$
DOCUMENT: Investment ($/yr^2)—The rate at which the infrastructure is enhanced or replenished.

$depreciation.N = GNP.N * (res\ deprec\ eff.N + gnpc\ deprec\ eff.N)$
$depreciation.S = GNP.S * (res\ deprec\ eff.S + gnpc\ deprec\ eff.S)$
DOCUMENT: Depreciation ($/yr^2)—The rate at which the economic infrastructure wears out or is used up.

$Population.N(t) = Population.N(t - dt) + (births.N - deaths.N) * dt$,
INIT $Population.N = 1000$
$Population.S(t) = Population.S(t - dt) + (births.S - deaths.S) * dt$,
INIT $Population.S = 2 * Population.N$
DOCUMENT: Population (population units)—Initial regional population.

$births.N = rgnpc\ birth\ eff.N * mortality.N * Population.N$
$births.S = rgnpc\ birth\ eff.S * mortality.S * Population.S$
DOCUMENT: Births (people/yr)—A flow representing the number of people who are born each year. For population to remain stable, this flow must equal the number who die. Thus, births is determined by the product of mortality rate and total population modified by the gross national product per capita effect.

$deaths.N = \max(0, Population.N * mortality.N)$
$deaths.S = \max(0, Population.S * mortality.S)$
DOCUMENT: Deaths (people/yr)—A draining flow representing the number of people who die each year.

$Resources.N(t) = Resources.N(t - dt) + (regeneration.N + resource\ trade - consumption.N) * dt$, INIT $Resources.N = 5e9$

$Resources.S(t) = Resources.S(t - dt) + (regeneration.S - consumption.S - resource\ trade) * dt$, INIT $Resources.S = 3 * Resources.N$

DOCUMENT: Resources (resource units)—The stock of resources available in the natural environment for extraction through technology. This value constitutes a complex aggregate of all raw materials that can be converted into finished products. Included among these is the empirical measure of this study, forestation.

$regeneration.N = \max(1.7 - norm\ Resources.N, 0) * Resources.N * regen\ rate$

$regeneration.S = \max(1.7 - norm\ Resources.S, 0) * Resources.S * regen\ rate$

DOCUMENT: Regeneration (resources/yr)—A compounding process is used to depict resource regeneration. The regeneration flow is defined as the product of resources and its regeneration rate. The max function works simply to limit the amount of resource regeneration should resources get too low to regulate resource regeneration.

$consumption.N = res\ cons\ eff.N * consum\ rate * GNP.N$
$consumption.S = res\ cons\ eff.S * consum\ rate * GNP.S$

DOCUMENT: Consumption (resources/yr)—A draining flow, the resources used each year driven by GNP.

$resource\ trade =$ if $(res\ trade\ eff > 0)$
then $\max(0, res\ trade\ eff * Resources.S)$
else $\max(0, res\ trade\ eff * Resources.N)$

DOCUMENT: Resource Trade (resource units/year)—The raw resources that flow between North and South.

$avg\ life\ span.N = 1/mortality.N$
$avg\ life\ span.S = 1/mortality.S$

DOCUMENT: Average Life Span (yrs)—Simply the inverse of the mortality rate.

$consum\ rate = 67.2$

DOCUMENT: Consumption Rate (res/$)—The amount of resources it takes to generate one $ of economic output or GNP. This value essentially balances the regeneration rate.

GNPC time delay = 20
DOCUMENT: Time Delay Gross National Product per Capita (yr)—The delay in years it takes for real GNP per capita changes to become incorporated into the national consciousness.

GNP growth.N = 100 ∗ (*GNP.N* − DELAY(*GNP.N*, 1))/DELAY(*GNP.N*, 1)
GNP growth.S = 100 ∗ (*GNP.S* − DELAY(*GNP.S*, 1))/DELAY(*GNP.S*, 1)
DOCUMENT: GNP growth rate.

GNP per capita.N = *GNP.N/Population.N*
GNP per capita.S = *GNP.S/Population.S*
DOCUMENT: GNP per capita ($/person-year)—Represents, per standard economics, the goods and services received by one person for one year.

Lateral Pressure = if (*trade switch*) then (
if (*GNP.N* > *GNP.S*)
then ((*GNP.N* − *GNP.S*)/(*GNP.N* + *GNP.S*))/*res cons eff.N*
else ((*GNP.N* − *GNP.S*)/(*GNP.N* + *GNP.S*))/*res cons eff.S*)
else 0
DOCUMENT: Lateral Pressure ()—Essentially driven by differences in economic strength between North and South, although resource consumption effects exacerbate Lateral Pressure through resource shortage and increased prices. Lateral Pressure, as used here, constitutes variable trade strength. Thus, it represents something in between the Heckscher-Ohlin trade model that assumes low factor specificity or high North/South transferability and the Ricardo-Viner model that assumes high factor specificity or low transferability.

norm GNP.N = *GNP.N*/INIT(*GNP.N*)
norm GNP.S = *GNP.S*/INIT(*GNP.S*)
DOCUMENT: Normalized GNP ()—A dimensionless ratio showing GNP, this study's proxy for technology, compared to its initial value.

norm Population.N = *Population.N*/INIT(*Population.N*)
norm Population.S = *Population.S*/INIT(*Population.S*)
DOCUMENT: Normalized Population ()—A dimensionless value denoting the relative size of the Population relative to its initial value.

norm Resources.N = Resources.N/init(Resources.N)
norm Resources.S = Resources.S/init(Resources.S)
DOCUMENT: Normalized Resources ()—A dimensionless value denoting the relative availability of resources.

population growth.N = 100 * (Population.N-delay(Population.N,1))/ delay(Population.N,1)
population growth.S = 100 * (Population.S-delay(Population.S,1))/ delay(Population.S,1)
DOCUMENT: population growth rate.

regen rate = .016
DOCUMENT: Regeneration Rate (%/yr)—Rate of resource regeneration. Modify with respect to consumption rate.

relative gnpc.N = min(3750, GNP per capita.N)/smth3(GNP per capita.N, GNPC time delay, GNP per capita.N)
relative gnpc.S = min(3750, GNP per capita.S)/smth3(GNP per capita.S, GNPC time delay, GNP per capita.S)
DOCUMENT: Relative GNP per capita ()—A normalized value representing the collective population's thought process regarding how well they're doing. If the value is greater than 1, then the population thinks it is doing well, and if the value is less than 1, then the population thinks it is doing poorly. This value changes over time as the denominator "catches up" to real GNP per capita in the numerator, a process controlled by GNPC time delay. The numerator is cut off at $3750/person-yr denoting that increases beyond this amount do not increase the birth rate just as increases beyond this amount do not decrease the death rate (cf. *mortality*).

resource change.N = 100 * (Resources.N − delay(Resources.N, 1))/ delay(Resources.N,1)
resource change.S = 100 * (Resources.S − delay(Resources.S, 1))/ delay(Resources.S,1)
DOCUMENT: percentage change in resources.

tech trade.N = if (*Lateral Pressure > 0*) then *developed* else *undeveloped*
DOCUMENT: Technical Trade North (%/yr)—If lateral pressure is positive, then northern countries are developed; if negative, then undeveloped.

tech trade.S = if (*Lateral Pressure* > 0) then *undeveloped* else *developed*
DOCUMENT: Technical Trade South (%/yr)—If lateral pressure is negative, then southern countries are developed; if positive, then undeveloped.

trade switch = 1
DOCUMENT: *trade switch* ()—simply turns trade effects on (1) and off (0).

developed = GRAPH(*Lateral Pressure*) (0.00, 0.00), (0.2, 0.00025), (0.4, 0.0005), (0.6, 0.00085), (0.8, 0.0011), (1, 0.00135), (1.20, 0.0017), (1.40, 0.0021), (1.60, 0.0027), (1.80, 0.0034), (2.00, 0.00435), (2.20, 0.00605), (2.40, 0.00795)
DOCUMENT: Developed (%/yr)—Graph translates from lateral pressure to investment effects though technical trade. Developed countries are the dominant trading partner. Since trade must influence a bigger economy, its effect is less than that for a developing economy. Lateral pressure and technical trade effects however yield positive influences for both partners, thus reflecting the "win-win" nature of comparative advantage.

gnpc deprec eff.N = GRAPH(*GNP per capita.N*) (0.00, 0.12), (625, 0.098), (1250, 0.09), (1875, 0.088), (2500, 0.084), (3125, 0.082), (3750, 0.08), (4375, 0.078), (5000, 0.076), (5625, 0.074), (6250, 0.072)
gnpc deprec eff.S = GRAPH(*GNP per capita.S*) (0.00, 0.12), (625, 0.098), (1250, 0.09), (1875, 0.088), (2500, 0.084), (3125, 0.082), (3750, 0.08), (4375, 0.078), (5000, 0.076), (5625, 0.074), (6250, 0.072)
DOCUMENT: Technology Depreciation Effect (%/yr)—As *GNP per capita* decreases, more pressure is placed on the common, public infrastructure, thus causing it to wear out faster.

invest rate.N = GRAPH(*GNP per capita.N*) (0.00, 0.00), (625, 0.08), (1250, 0.09), (1875, 0.095), (2500, 0.1), (3125, 0.105), (3750, 0.109), (4375, 0.113), (5000, 0.116), (5625, 0.119), (6250, 0.12)
invest rate.S = GRAPH(*GNP per capita.S*) (0.00, 0.00), (625, 0.08), (1250, 0.09), (1875, 0.095), (2500, 0.1), (3125, 0.105), (3750, 0.109), (4375, 0.113), (5000, 0.116), (5625, 0.119), (6250, 0.12)
DOCUMENT: Investment Rate (%/yr)—The percentage of GNP that is reinvested back into the infrastructure each year. Investment is, of course, related to savings, which is northern, industrialized countries is around 10% per year (Samuelson and Nordhaus 1992, 445). The curve

is nonlinear, indicating that investment falls off fairly linearly until the economy completely collapses at low *GNP per capita* levels. Finally, note that the difference between investment rate and depreciation rate yields the growth rate of total output or GNP.

mortality.N = GRAPH(*GNP per capita.N*) (0.00, 0.1), (312, 0.07), (625, 0.05), (938, 0.037), (1250, 0.03), (1562, 0.027), (1875, 0.024), (2188, 0.022), (2500, 0.02), (2812, 0.018), (3125, 0.016), (3438, 0.015), (3750, 0.014)

mortality.S = GRAPH(*GNP per capita.S*) (0.00, 0.1), (312, 0.07), (625, 0.05), (938, 0.037), (1250, 0.03), (1562, 0.027), (1875, 0.024), (2188, 0.022), (2500, 0.02), (2812, 0.018), (3125, 0.016), (3438, 0.015), (3750, 0.014)

DOCUMENT: Mortality (%/yr)—Fraction of the population that dies each year, which is assumed to depend on *GNP per capita*. The higher the GNP per capita, the greater the average life-span due to improvements in nutrition and medical care among other factors. Note that the inverse of mortality is life span, e.g., $1/(2\%/\text{yr}) = 50$ years

res cons eff.N = GRAPH(*norm Resources.N*) (0.00, 0.00), (0.2, 0.4), (0.4, 0.64), (0.6, 0.8), (0.8, 0.92), (1.00, 1.00), (1.20, 1.07), (1.40, 1.11), (1.60, 1.15), (1.80, 1.18), (2.00, 1.20)

res cons eff.S = GRAPH(*norm Resources.S*) (0.00, 0.00), (0.2, 0.4), (0.4, 0.64), (0.6, 0.8), (0.8, 0.92), (1.00, 1.00), (1.20, 1.07), (1.40, 1.11), (1.60, 1.15), (1.80, 1.18), (2.00, 1.20)

DOCUMENT: Resource Consumption Effect ()—Dimensionless value that modifies Consumption Rate. Captures the dynamic that as the amount of available resources declines, they become more expensive and are thus used more efficiently. At the extreme, should resources ever reach zero, so will consumption.

res deprec eff.N = GRAPH(*norm Resources.N*) (0.00, 0.2), (0.05, 0.14), (0.1, 0.093), (0.15, 0.06), (0.2, 0.037), (0.25, 0.024), (0.3, 0.016), (0.35, 0.01), (0.4, 0.005), (0.45, 0.002), (0.5, 0.00)

res deprec eff.S = GRAPH(*norm Resources.S*) (0.00, 0.2), (0.05, 0.14), (0.1, 0.093), (0.15, 0.06), (0.2, 0.037), (0.25, 0.024), (0.3, 0.016), (0.35, 0.01), (0.4, 0.005), (0.45, 0.002), (0.5, 0.00)

DOCUMENT: Resource Depreciation Effect (%/yr)—The increase in depreciation that occurs as resources become scarce, thus (1) raw materials become more expensive, and (2) preventive maintenance gets deferred causing investments to wear out faster.

res trade eff = GRAPH(*Lateral Pressure*) (−2.40, −0.4), (−2.00, −0.272), (−1.60, −0.16), (−1.20, −0.08), (−0.8, −0.028), (−0.4, −0.006), (0.00, 0.00), (0.4, 0.006), (0.8, 0.028), (1.20, 0.08), (1.60, 0.16), (2.00, 0.272), (2.40, 0.4)

DOCUMENT: Resource Trade Effect (%/yr)—Converts from lateral pressure value to its effect on the trade of resources from North to South.

rgnpc birth eff.N = GRAPH(*relative gnpc.N*) (0.00, 0.00), (0.2, 0.025), (0.4, 0.075), (0.6, 0.15), (0.8, 0.4), (1.00, 1.00), (1.20, 2.40), (1.40, 3.62), (1.60, 4.50), (1.80, 4.93), (2.00, 5.00)

rgnpc birth eff.S = GRAPH(*relative gnpc.S*) (0.00, 0.00), (0.2, 0.025), (0.4, 0.075), (0.6, 0.15), (0.8, 0.4), (1.00, 1.00), (1.20, 2.40), (1.40, 3.62), (1.60, 4.50), (1.80, 4.93), (2.00, 5.00)

DOCUMENT: Relative GNP per Capita Birth Effect ()—A modifier. Basically, the higher the GNP per capita, the higher the birth rate. However, this effect tends to diminish over time. This dynamic is captured by the GNPC set point. So if people think they are doing well economically, GNPC set point is greater than 1; if they think they are doing poorly, it is less than 1; and if they are doing satisfactorily, then the value is 1. Note that when the input and output values are both 1, then the birth rate equals death rate, and population growth is 0.

undeveloped = GRAPH(*Lateral Pressure*) (0.00, 0.00), (0.2, 0.001), (0.4, 0.0025), (0.6, 0.004), (0.8, 0.006), (1, 0.00825), (1.20, 0.0113), (1.40, 0.014), (1.60, 0.0173), (1.80, 0.022), (2.00, 0.0278), (2.20, 0.0365), (2.40, 0.0498)

DOCUMENT: Undeveloped (%/yr)—Graph translates from lateral pressure to investment effects though technical trade. Undeveloped countries are the weaker trading partner. Since trade influences a smaller economy, its effect is greater than that for a developed economy. Lateral pressure and technical trade effects however yield positive influences for both partners, thus reflecting the "win-win" nature of comparative advantage.

Appendix F
Nonlinear Relationship Analysis

The nonlinear, graphical, or tabular relationships—initially presented throughout appendix E—are examined here more visually and thoroughly. Besides simply showing what the nonlinear relationships (as addressed initially in figure 1.1) look like, the reasoning behind the curve's shape is explicitly presented. Note that all curves, while nonlinear, increase or decrease *montonically*. That is, if the curve decreases, it will never reverse direction and increase. Conversely, if the curve increases, it will never reverse direction and decrease. Curves that both increase and decrease indicate an attempt to place too much explanatory power into too few variables. Such situations must be thought through more carefully, and more variables must be added to the model to account for the underlying complexity.

The converters *death fraction* and *consumption per day* are nonlinear tables as defined by figure F.1. The converter's input is the variable on the horizontal or X axis, and the output is the variable on the vertical or Y axis. The graph showing death fraction's output goes up markedly as its input, *food per capita,* decreases. The graph can be interpreted as follows: As the amount of available food per person increases, the death rate decreases (other things being equal), although additional food eventually has no further affect on the death rate. Conversely, as food per capita increases, so does consumption per day. This makes sense because as the amount of food increases, people tend to eat more, although this effect also diminishes after food per capita reaches a certain point.

Population, as with the overshoot and collapse model, is modified by two flows, *deaths* and *births*. The deaths flow is driven by the graphical converter *mortality,* which is determined by *GNP per capita*. Its graphical

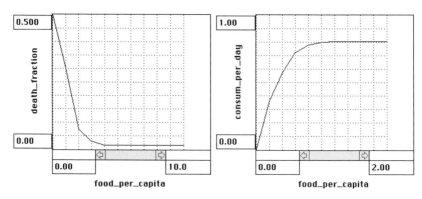

Figure F.1
Overshoot and Collapse Converter Graphs.
output: *death fraction* output: *consumption per day*
input: *food per capita* input: *food per capita*

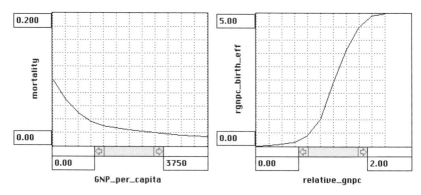

Figure F.2
Population Converter Graphs.
output: *mortality* output: *relative GNP per capita birth effect*
input: *GNP per capita* input: *relative GNP per capita*

function is included in figure F.2 and shows that mortality falls as GNP per capita increases due to improvements in diet and medical care. The births flow is defined as the product of mortality, population, and the graphical converter *rgnpc birth effect*. This makes sense because stable populations require that the number of births equal the number of deaths. The value of rgnpc birth effect is determined by *relative gnpc*, which gives a sense of how people think they are doing economically: If they think they are doing well, relative gnpc is greater than 1; poorly, less than 1; getting by, 1. Divergence from unity, the neutral value, tends to diminish as people become accustomed to their current living conditions. These expectations are reflected in rgnpc birth effect—as economic expectations increase, so does the birth rate. Consequently, an increase in *GNP* leads to a decrease in deaths and a temporary increase in births causing population to grow until the birth rate drops to match the death rate.

The *GNP* stock is modified by two flows, *investment* and *depreciation*. Figure F.3 shows that as GNP per capita increases, so does investment rate. The investment rate is based on the saving rate, which is approximately 10 percent for developed countries (Samuelson and Nordhaus 1992, 445). The depreciation flow, which is partially determined by *gnpc deprec effect*, demonstrates the opposite relationship: as GNP per capita increases, depreciation decreases. This is done on the assumption that as GNP/capita falls, individuals push more of their personal costs onto the public sector causing more wear and tear on the nation's infrastructure. Growth of *GNP* is determined by the difference between investment and depreciation, which has run about 3 percent for the United States since World War II (Samuelson and Nordhaus 1992, 399). These relationships are all incorporated into the graphical functions of figure F.3.

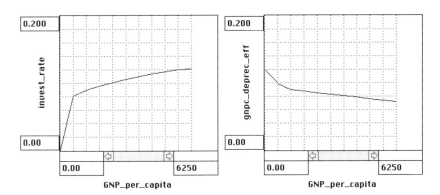

Figure F.3
GNP Converter Graphs.
output: *investment rate*   output: *GNP per capita depreciation effect*
input: *GNP per capita*   input: *GNP per capita*

The *Resources* stock is modified by the *consumption* and *regeneration* flows. The regeneration flow represents the rate at which natural resources regenerate themselves, which can best be envisioned in terms of the study's empirical measure, forest area. If a patch of forest is chopped down, it recovers or regenerates at a rate measured in terms of forest area/year. The regeneration flow is actually measured in terms of resources/year. The consumption flow represents the rate at which natural resources are consumed, a process driven by *GNP*. Note that the flow aspect of GNP, measured in terms of $/year, is made explicit as a growing GNP requires an increased flow of natural resources to support that growth. Figure F.4 shows how the system reacts to depleting resources. The left graph demonstrates the relationship between *resource conservation effect* and *normalized Resources,* a percentage created by dividing the current value of Resources by its initial value. Essentially this graph depicts how resources are used more efficiently as they become both less abundant and more expensive, the term *efficiency* implying that fewer natural resources are used to generate the same level of GNP. The graph to the right shows the relationship between *resource depreciation effect* and normalized Resources. Here as resources become rarer and more expensive, fewer raw materials are available to maintain society's infrastructure thereby increasing depreciation.

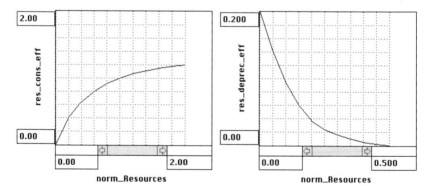

Figure F.4
Resource Converter Graphs.
output: *resource conservation effect*   output: *resource depreciation effect*
input: *normalized Resources*   input: *normalized Resources*

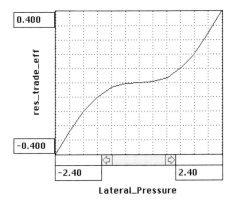

Figure F.5
Resource Trade Effect Converter Graph.
output: *resource trade effect*
input: *Lateral Pressure*

The freedom with which resources flow between North and South is determined by *resource trade effect*, a function of *Lateral Pressure*. Lateral Pressure is driven by the relative weights of northern and southern GNP: if *GNP.N* is greater than *GNP.S*, then Lateral Pressure is positive; if GNP.S is greater than GNP.N, then Lateral Pressure is negative. Note that because the initial values of GNP.N and GNP.S are equal, the initial value of Lateral Pressure is 0. Lateral Pressure then drives resource trade effect as shown in figure F.5. The output of resource trade effect represents the percentage of the Resources stock that is transferred between North and South in one year. It is revealing to note how this trade model relates to more established economic theories. First, diverging economic output values contribute to the production diversity that drives comparative advantage. If one country or region has a high GNP, then it will have more incentive to trade with low GNP regions and vice versa. The second revealing aspect of the ELP trade relationship regards *factor specificity*, the ease or difficulty with which commodities or products move among regions. Previous theories have assumed one extreme of factor specificity or the other: the Heckscher-Ohlin trade model assumes low factor specificity or high resource transferability among disparate geographic regions, while the Ricardo-Viner trade model assumes high factor specificity or low resource transferability (Alt et al. 1996). Lateral Pressure, as used here, provides the means to vary factor specificity endogenously, that is, within the analytical confines of the ELP model. When Lateral Pressure is close to zero, factor specificity is high; when it is strongly positive or negative, factor specificity is low.

High technology trade, of the type that flows from North to South, is treated somewhat differently than the resource-based trade that flows from South to North. Figure 5.5 contains a physical flow connecting Resources.N to Resources.S, denoting that if a resource flows from South to North, then it is no longer available for use in the South because it has been physically transferred to the North. Technology-based trade is not quite so simple because if a technology is transferred from the North to the South, then that technology is still available for use in the North. For example, if the United States shares its software with a developing nation, then the United States still retains its ability to use that software. Conversely, if the developing country cuts down all its trees and sells them to the United States to pay for the software—then those trees are gone, at least until new ones slowly grow to replace them. The dynamics of technical trade are captured by the graphs in figure F.6, which yield growth rates that are added to the investment flows of the North and South as shown in figure 5.6. Per the tenets of comparative advantage, both developed

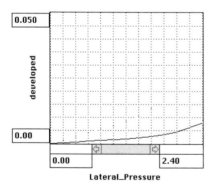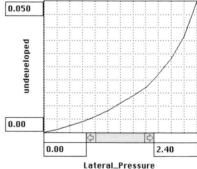

Figure F.6
Technical Trade Converter Graphs.
output: *developed* growth contribution of technical trade
input: *Lateral Pressure*

output: *undeveloped* growth contribution of technical trade
input: *Lateral Pressure*

and undeveloped countries benefit from trade. In fact, undeveloped countries benefit about five times more than developed countries according to figure F.6. These graphs are scaled this way because undeveloped nations tend to have smaller GNPs, and so an infusion of high-technology goods will generate a larger percentage impact. Conversely, developed nations tend to have large GNPs, and so the additional profits from technical trade tend to pack a somewhat smaller economic punch.

Appendix G
Dynamic Equilibrium Analysis

In creating a system dynamics model, establishing an analytical baseline is an important part of the analysis because from there, causality can be attributed. In other words, one can then determine linkages between causes and effects. Conversely, if the model gyrates and generate results all the time, then how can one determine causality? In chapter 5, recall that there were two time-frames in question: before World War II without trade, and after World War II with trade. Also, there were two regions of interest, the developed North and developing South. These cleavages lead to four permutations: (1) North without trade, (2) North with trade, (3) South without trade, and (4) South without trade. The first three are addressed in section 5.3.3, the fourth is addressed here.

While the dynamic response of figure G.1 may at first glance appear uninteresting, it is important for several reasons. First, flat southern dynamics without trade form an important part of the larger argument that the North pushes its environmental costs onto the South through trade. Without trade, the South exists in a sustainable manner with population, technology, and resources remaining in equilibrium. It might be tempting to conclude that "nothing is happening" in this simulation, but the feedbacks are all operating, the flows are flowing, and the calculations are calculating. The simulation shows that the stocks do not vary because the flows into and out of each are equal. This steady state response thus constitutes both a *dynamic equilibrium* (Nicolis and Prigogine 1989, 54–56) and an example of what "sustainability" looks like. The system is "in motion" yet remains viable indefinitely.

Second, figure G.1 provides a methodological as well as an analytical baseline so that causality can be correctly attributed when the growth

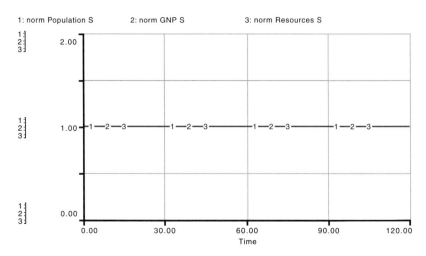

Figure G.1
South Dynamics without Trade. This response has three interpretations. First, from a physics perspective, it is an example of "dynamic equilibrium." Second, from an environmental politics perspective, it is an example of, "sustainability." Third, from a methodological perspective, it provides the analytic baseline for the dynamic analysis presented in section 5.3.3.

and trade processes of section 5.3.3 are incorporated into the model. For example, when the stocks are initialized to a northern orientation, the resulting dynamic changes can be attributed to the new initial conditions (cf. figure 5.10). And when trade connections are established between North and South, then the resulting changes in the South can be attributed to trade (cf. figure 5.12). In this way, the analysis proceeds from South to North, both without trade—the two scenarios differing only in their initial conditions, which must therefore be the cause of their dynamic differences. Modern trade relationships then connects the two regions, which in turn drive additional system behaviors and dynamics.

Notes

1. For a more detailed discussion of these trade disputes, see the beginning of chapter 3.

2. The World Economic Forum meets every year in Davos, Switzerland, and has become a place where world leaders go to discuss global economic issues. The meeting was started by Klaus Schwab, a professor of business policy at the University of Geneva, and it has become remarkably popular since its inception in 1970.

3. The "new geopolitics" actually refers to three separate trends: first, to spatial-analytic, computer-based geopolitics as previously noted; second, to geopolitics as geoeconomics, see for example, Agnew 1995; third, to critical geopolitics, for example, O'Tuathail 1996.

4. For a discussion of microfeatures and macrobehavior in terms of fractals and chaos, see appendix A.

5. The feedback concept is developed more explicitly in section 5.2 and is used generally throughout chapter 5.

6. Physicist Richard Feynman, in a 1964 lecture at Cornell University, explained the way to look for a new scientific law: (1) guess it, (2) compute the consequences of the guess, (3) compare the guess with nature by experiment, experience, or observation. If the guess disagrees with experiment, then it is wrong. This relationship between guess and experiment is the key to science (Sykes 1993).

7. An "image" consists of the philosophical combination of space, time, personal relationships, physical laws, and more subtle intimations and emotions. In short, an image represents the knowledge or world view that influences an individual's actions (Boulding 1956, 3–6).

8. The relationship between system structure and dynamics is implicit within the organization of chapter 5. Essentially, the two are connected through feedback, the concept of circular causal connections. For more on feedback thought in social systems, see Richardson 1991.

9. The additional respondents within Keohane (1986a) are John Gerard Ruggie, Robert W. Cox, Richard K. Ashley, and Robert G. Gilpin.

10. Simon's bounded rationality and its relationship to lateral pressure is explicitly addressed in section 2.5.

11. Morgenthau (1985 [1948], chap. 9) lists the following elements of power: geography, natural resources (food, raw materials), industrial capacity, military preparedness (technology, leadership, quantity and quality of armed forces), population (distribution, trends), national character, national morale, the quality of diplomacy, the quality of government. This list is not definitive—others would provide different power element lists.

12. To delve more deeply into the tension between environment and individual, see the opportunity and willingness framework developed by Most and Starr (1989), and the ecological triad model of entity, environment, entity/environment interaction developed by Sprout and Sprout (1965, 1968, 1971).

13. The relationship between Simon's bounded rationality and the methodological underpinnings of lateral pressure is addressed further in section 2.5, there the more technical aspects lateral pressure and system are developed.

14. Choucri and North 1989 contains preliminary findings from Choucri, North, and Yamakage 1992 as well as a direct comparison with Choucri and North 1975.

15. A great deal of environment and politics work has been done with respect to the atmospheric issues of CO_2, CFCs, and global warming. This study uses deforestation as its indicator to capture causal connections among economic activity, land change, and environmental degradation.

16. It is interesting to note that the words *polity, policy,* and *police* all derive from the same Latin root *politia* meaning government or administration. The modern derivatives all relate to matters of effective governance, rules, and maintenance of morality.

17. Similarly, the academy reflects this division of the social and natural through the separate social and natural sciences, each institutionally insulated from the other (Choucri 1993b, 9).

18. Forest change and deforestation are explored further within sections 3.1 and 3.2.7.

19. The discussion that follows contrasts and compares the quantitative methods used in this study in a general manner. Consequently, some familiarity with quantitative methodology is assumed. Note, however, that each method is developed in more detail throughout the remaining analytical chapters.

20. The overall organization of this study is based on the system dynamics' reference mode, which helps to clarify expectations and focus research effort (Randers 1980, 121–127). The reference mode concept, as used herein, consists of five parts: (1) identify the timeframe and major themes of the study, (2) describe and develop the theory that motivates and organizes the study, (3) identify primary variables and graph their dynamics, (4) identify sign and magnitude of causal connections among the variables, and (5) develop a system dynamics model based on the previously developed timeframe, themes, theory, variables, dynamics, and connections.

21. The term "diffusion" is the value neutral analogue to the more value-laden term "contagion." The terms are effectively interchangeable as both address geographically dependent spreading or expanding processes.

22. Feedback in simultaneous equations implies that if $S = f(N)$, then $N = f(S)$. Taking N to represent the natural environment and S the social, the social environment is a function of the natural and vice versa. This "circular" relationship can be extended to an arbitrary number of variables and equations so long as certain mathematical and data constraints are satisfied (Maddala 1992, 373–377).

23. Rucker (1987, 290–301) makes this point from the perspective of twentieth-century mathematics by explicitly contrasting the complexity of the world with that of the human mind in a discussion entitled "Inconceivability."

24. Migration, as denoted by the dashed line in figure 2.4, is also a form of lateral pressure. Owing to the complexity of global-scale population issues, migration is not explicitly developed in this analysis. However, several relevant works point the way into the literature from the lateral pressure perspective. Choucri (1984) looks at population and conflict, and Demeny (1990) addresses population from the perspective of geography and the environment. Perhaps most evocative is the view of Kennedy (1996, 58) who sees modern migration as a consequence of the industrial revolution's diffusion from the British Isles. This observation squares nicely with that of Turner et al. (1990), who also view global environmental degradation as a consequence of the industrial revolution.

25. Choucri 1993b uses CO_2 as an environmental indicator because atmospheric effluents, being free to cross national borders, contribute to environmental lateral pressure. This study, in contrast, evaluates environmental lateral pressure through the impact of international trade processes on the natural environment as measured by forest change and deforestation.

26. This dynamic repeated itself in 1998 when the WTO dispute settlement panel found for India, Malaysia, Pakistan, and Thailand against the United States regarding shrimp and sea turtles. In this case, the United States halted shrimp imports from these developing countries because too many turtles were being killed in the nets used to catch the shrimp. The WTO ruled, as did GATT before it, that countries must treat like products similarly regardless of their method of production, which includes environmental consequences (Lane 1998a).

27. North and Choucri 1996 reverses groups 3 and 4 so that group 3 denotes *area* > *gnp* > *pop* and group 4, *pop* > *gnp* > *area*. This more recent definition is rejected here because the original profile definitions were symmetric with respect to population and technology: for example, *pop* > *gnp* held for groups 1–3 and *gnp* > *pop* for groups 4–6. This symmetry proves useful for the subsequent analysis, and so the original Choucri and North (1993b, 73) definitions are retained.

28. The following ecological groupings were generated from the definitions in table 3.1 and the 1991 data from World Bank 1995. There are 156 total countries, and values in parentheses denote group subtotals—Group 1 (25): Algeria, Argentina, Belize, Bolivia, Botswana, Central African Republic, Chad, Congo,

Czech Republic, Equatorial Guinea, Gabon, Guyana, Kazakhstan, Mali, Mauritania, Namibia, Niger, Papua New Guinea, Paraguay, Russian Federation, Slovak Rep., Solomon Islands, Turkmenistan, Vanuatu, Zambia; Group 2 (47): Benin, Bhutan, Brazil, Burkina Faso, Burundi, Cambodia, Cameroon, Chile, China, Colombia, Cote d'Ivoire, Ecuador, Egypt, Ethiopia, Gambia, Ghana, Guinea, Guinea-Bissau, Honduras, Indonesia, Jordan, Kenya, Kyrgyz Republic, Laos, Lesotho, Madagascar, Malawi, Morocco, Mozambique, Nepal, Nicaragua, Nigeria, Pakistan, Peru, Sao Tome and Principe, Senegal, Sierra Leone, Swaziland, Tajikistan, Tanzania, Togo, Uganda, Uruguay, Uzbekistan, Venezuela, Western Samoa, Zimbabwe; Group 3 (46): Armenia, Azerbaijan, Bangladesh, Belarus, Bulgaria, Cape Verde, Comoros, Costa Rica, Dominica, Dominican Republic, El Salvador, Estonia, Fiji, Georgia, Grenada, Guatemala, Haiti, Hungary, India, Iran, Jamaica, Kiribati, Latvia, Lithuania, Malaysia, Maldives, Mauritius, Mexico, Moldova, Panama, Philippines, Poland, Romania, Rwanda, South Africa, Sri Lanka, St. Kitts and Nevis, St. Lucia, St. Vincent and the Grenadines, Syria, Thailand, Tonga, Trinidad and Tobago, Tunisia, Turkey, Ukraine; Group 4 (6): Australia, Canada, Iceland, Oman, Saudi Arabia, Suriname; Group 5 (3): Finland, New Zealand, Norway; Group 6 (29): Antigua and Barbuda, Austria, Bahamas, Bahrain, Barbados, Cyprus, Denmark, France, Germany, Greece, Hong Kong, Ireland, Israel, Italy, Japan, Malta, Netherlands, Portugal, Puerto Rico, Qatar, Seychelles, Singapore, South Korea, Spain, Sweden, Switzerland, United Arab Emirates, United Kingdom, United States.

29. Six elements constitute 95 percent of the Earth's biomass—the material that makes up plants, animals, and microbes—which makes these elements the most critical to the planet's biogeochemical cycles. They are hydrogen, oxygen, nitrogen, carbon, sulfur, and phosphorus (Chiras 1989, 84).

30. For a work that considers multiple social and natural variables over the past three hundred years, see Turner et al. 1990.

31. This is not quite as long as the time scale envisioned by North and Choucri (1996, 2), from the end of the World War II (1945) to the present, but given the paucity of quality environmental time-series data, World Bank 1995 provides a good basis for an initial effort.

32. In discussing GNP per capita, it should be noted that there exists disagreement regarding the measure's meaning. GNP per capita is usually cited as a measure of economic welfare, with GNP or total output representing a stock of wealth to be dispersed among the citizenry to satisfy a range of needs and wants. Daly 1991 argues instead that GNP is essentially a yearly cost or flow required to maintain a nation's stock of wealth. That said, this analysis treats GNP per capita as a measure of welfare while acknowledging the measure's shortcomings.

33. North and South totals do not add up to world totals in figure 3.8 because many of the former Eastern Bloc nations are excluded from the profile analysis as their GNP data are unavailable. The CO_2 outputs for the former Eastern Bloc nations are therefore included in the world totals but are not included in the North or South totals.

34. These large migration values were removed from the analysis to provide a truer picture of long-term, birth-based population changes.

35. It is interesting to recall that the word *tradeoff* first appeared in English dictionaries in 1962, during the "halcyon, golden days of Keynesianism" (Ward, Davis, and Lofdahl 1995, 33).

36. This argument is directed solely at the relative balance between costs and revenues, the very revenues Bhagwati (1993) maintains will be spent correcting the costs. Whether it is reasonable to expect that present social institutions would actually direct revenues toward environmental problems assuming that unspent revenues and technologically feasible solutions actually exist is left as an open question.

37. The G_i^* equation is defined per the equation

$$G_i^* = \frac{\sum_{j=1}^{n} w_{ij}(d)x_j - W_i^*\bar{x}}{s\left\{[(nS_{1i}^*) - W_i^{*2}]/(n-1)\right\}^{1/2}}.$$

38. The statistics and graphics were generated using the General Linear Model (GLM) function from the S-PLUS™ statistical package (version 3.3) running on a Sun SPARC™ 5 workstation with the Sun Solaris™ 2.4 operating system, a variation of UNIX™.

39. In interpreting this table, and the others like it that follow, note that $\hat{\beta}_1$ denotes the parameter estimate for the first coefficient that corresponds to the model's first independent variable. The first estimate, $\hat{\beta}_0$, corresponds to the intercept. The product of parameter estimates and mean values provides an intuitive way to evaluate the magnitude of the estimates. The values can be compared directly to the dependent variable since they have the same units. Ratios of the parameter products with respect to the mean dependent variable are provided as a rough, order of magnitude way to gauge the estimate's influence within the model.

40. It might be questioned whether CO_2 and GNP are strictly comparable values, CO_2 being a flow of effluent per year and GNP being a stock of wealth to be spent on personal wants and needs. GNP is not viewed as such in this analysis. Instead, it is viewed as the flow of goods, materials, and resources needed to maintain a country's infrastructure for one year (Daly 1991, 102–103).

41. Forest change = f(Agricultural land change): $\hat{\beta}_1$, −0.0211; σ, 0.0034; t, −6.1986; pr, 0.0000.

42. When tested with respect to absolute latitude, TC × GNP exhibits a negative sign as did deforestation, which indicates that as one moves away from the poles, TC × GNP increases. Full absolute latitude results are as follows: $\hat{\beta}_{|lat|}$, −0.0056; σ, 0.0029; t, −1.92; pr, 0.0568. When tested with respect to absolute longitude, TC × GNP exhibits a strong positive sign as did deforestation, which indicates that as one moves away from 0° longitude, TC × GNP increases. Full absolute longitude results follow: $\hat{\beta}_{|long|}$, 0.0055; σ, 0.0011; t, 4.95; pr, 0.0000.

43. The trade-based contiguity matrix is based on the import data from International Monetary Fund 1994, reworked into the 207 country format of World Bank 1995. The same tests were run on export-based trade matrices, but due to

the high correlation between imports and exports (cf. figure B.6), the results were essentially the same. The term *matrices* is used instead of *matrix* because import values change from year to year, and so a new contiguity matrix was generated for each year in the analysis.

44. Hardin (1968), arguing from more modern, microeconomic foundations, similarly concludes that the advances required to solve environmental degradation are not essentially technical but moral.

45. Meadows et al. 1972 is based on the more technical treatment by Forrester (1973) and is revisited by Meadows, Meadows, and Randers (1992).

46. Section 1.3 generally discusses complex social systems that can be represented by system dynamics models. Whereas section 1.3 only alludes to system dynamics, here it is addressed specifically.

47. Goodman (1974) provides many helpful examples of translating system dynamics models into causal-loop diagrams.

48. The full set of equations for the ELP model are included in section E.2. Additional documentation for and explanation of the model's individual variables can also be found there.

49. Note that North and South, as represented here, consist of the relationships previously developed in figures 5.4 and 5.8 and then connected in figures 5.5 and 5.6.

50. The only difference between this simulation run and that of figure G.1 that represents the South without trade, regards the model's initial conditions: Population.N has been halved to 1000 population units, and Resources.N has been reduced from 15×10^9 to 5×10^9 resource units. Northern and southern *GNPs* remain the same.

51. In 1990, southern countries defined as those with a greater global share of population than GNP, had twice the land area of northern countries defined as those with greater global share of GNP than population (Choucri and North 1993b; World Bank 1995). Because southern countries tend to be in warm, tropical regions as opposed to cool, temperate ones, and because forest area is the resource measure of this study, Resources.S is set to be three times Resources.N.

52. A fourth unique loop, formed by the migration connection, is pointed out for informational purposes but remains undeveloped in this study.

53. It is instructive to evaluate the exaggerated claims and scientific pretensions of Victorian era social reformers, legal positivists, and political philosophers in light of the more modest and historically verifiable claims described herein (cf. Fuller 1987).

54. Kristof 1999 is a four-part series for the *New York Times* consisting of Kristof and Wyatt 1999, Kristof and Sanger 1999, and Kristof and WuDunn 1999a, 1999b.

55. Ward and Lofdahl 1995 explore integration processes from a complex systems perspective using the case of the European Union.

Bibliography

Agnew, J. A. 1995. *Mastering space: Hegemony, territory and international political economy.* New York: Routledge.

Akira, S. 1989. *Capital accumulation in Thailand.* Tokyo: The Centre for East Asian Cultural Studies.

Alt, J. E., J. Frieden, M. J. Gilligan, D. Rodrik, and R. Rogowski. 1996. The political economy of international trade: Enduring puzzles and an agenda for inquiry. *Comparative political studies* 29(6): 689–717.

Amin, S. 1997. *Capitalism in the age of globalization.* London: Zed Books.

Aron, R. 1966. *Peace and war: A theory of international relations.* Garden City, NY: Doubleday.

Ashley, R. K. 1980. *The political economy of war and peace: The Sino–Soviet–American triangle and the modern security problematique.* New York: Nichols Publishing.

Axelrod, R. 1984. *The evolution of cooperation.* New York: Basic Books.

Axelrod, R., and M. D. Cohen. 1999. *Harnessing complexity: Organizational implications of a scientific frontier.* New York: Free Press.

Baker, D., R. Pollin, and G. A. Epstein. 1999. *Globalization and progressive economic policy: The real constraints and options.* Cambridge, England: Cambridge University Press.

Barnet, R. J., and J. Cavanagh. 1995. *Global dreams: Imperial corporations and the new world order.* New York: Simon and Schuster.

Beck, N., and J. N. Katz. 1995. How not to be misled by time-series–cross-section data: OLS vs. GLS-ARMA. (Unpublished manuscript.)

Bennett, J. W., and K. A. Dahlberg. 1990. Institutions, social organization, and cultural values. In *The earth as transformed by human action,* ed. B. L. Turner II, W. C. Clark, R. W. Kates, J. F. Richards, J. T. Mathews, and W. B. Meyer, 69–86. New York: Cambridge University Press.

Bhagwati, J. 1993. The case for free trade. *Scientific American* 269 (November): 42–49.

Bishop, C. M. 1995. *Neural networks for pattern recognition*. New York: Oxford University Press.

Boulding, K. E. 1956. *The image: Knowledge and life in society*. Ann Arbor: University of Michigan Press.

Bradley, E. 1995. Causes and effects of chaos. *Computers and graphics* 19(5): 755–778.

Braudel, F. 1979. *The perspective of the world: Civilization and capitalism, 15th–18th century*, Volume 3. New York: Harper & Row.

Brecher, J., and T. Costello. 1998. *Global village or global pillage*. Cambridge, MA: South End.

Brzezinski, Z. 1993. *Out of control: Global turmoil on the eve of the 21st century*. New York: Charles Scribner's Sons.

Burtless, G., R. Z. Lawrence, R. E. Latin, and R. J. Shapiro. 1998. *Globaphobia: Confronting fears about open trade*. Washington, DC: Brookings Institute.

Carr, E. H. 1964 [1939]. *The twenty years' crisis, 1919–1939*. New York: Harper & Row.

Cassidy, J. 1996. The decline of economics. *The New Yorker* (December): 50–60.

Cavanagh, J. 1992. *Trading freedom: How free trade affect our lives, work, and the environment*. San Francisco: Institute for Food and Development Policy.

Chiras, D. D. 1989. *Environmental science: A framework for decision making*. Menlo Park, CA: Addison-Wesley.

Chisholm, M. 1990. The increasing separation of production and consumption. In *The earth as transformed by human action*, ed. B. L. Turner II, W. C. Clark, R. W. Kates, J. F. Richards, J. T. Mathews, and W. B. Meyer, 87–102. New York: Cambridge University Press.

Choucri, N. 1978. System dynamics forecasting in international relations. In *Forecasting in international relations: Theory, methods, problems, prospects*, ed. N. Choucri and T. W. Robinson. San Francisco: W. H. Freeman. (Written with the assistance of Brian Pollins.)

Choucri, N. 1981. *International energy futures: Petroleum prices, power, and payments*. Cambridge, MA: MIT Press. (Written with David Scott Ross and the collaboration of Brian Pollins.)

Choucri, N. 1984. *Multidisciplinary perspectives on population and conflict*. Syracuse, NY: Syracuse University Press.

Choucri, N. (ed.) 1993a. *Global accord: Environmental challenges and international responses*. Cambridge, MA: MIT Press.

Choucri, N. 1993b. Introduction: Theoretical, empirical, and policy perspectives. In *Global accord: Environmental challenges and international responses*, ed. N. Choucri, 1–40. Cambridge, MA: MIT Press.

Choucri, N. 1993c. Multinational corporations and the global environment. In *Global accord: Environmental challenges and international responses*, ed. N. Choucri, 205–253. Cambridge, MA: MIT Press.

Choucri, N., and R. C. North. 1975. *Nations in conflict: National growth and international violence.* San Francisco: W. H. Freeman.

Choucri, N., and R. C. North. 1989. Lateral pressure in international relations: Concept and theory. In *Handbook of war studies,* ed. M. I. Midlarsky, 289–326. Ann Arbor: University of Michigan Press.

Choucri, N., and R. C. North. 1993a. Global accord: Imperatives for the twenty-first century. In *Global accord: Environmental challenges and international responses,* ed. N. Choucri, 477–507. Cambridge, MA: MIT Press.

Choucri, N., and R. C. North. 1993b. Growth, development, and environmental sustainability: Profile and paradox. In *Global accord: Environmental challenges and international responses,* ed. N. Choucri, 67–132. Cambridge, MA: MIT Press.

Choucri, N., R. C. North, and S. Yamakage. 1992. *The challenge of Japan before World War II and after: A study of national growth and expansion.* London: Routledge.

Cliff, A., and J. Ord. 1981. *Spatial processes: Models and applications.* London: Pion Limited.

Cole, S. 1977. *Global models and the international economic order.* New York: Permagon Press.

Cropsey, J. 1987a. Adam Smith. In *History of political philosophy* (3d ed.), ed. L. Strauss and J. Cropsey, 635–658. Chicago: University of Chicago Press.

Cropsey, J. 1987b. Karl Marx. In *History of political philosophy* (3d ed.), ed. L. Strauss and J. Cropsey, 802–828. Chicago: University of Chicago Press.

Cropsey, J. 1990. The nature of contemporary political science: A roundtable discussion. *PS: Political Science and Politics* 23(1): 34–43.

Daly, H. 1991. *Steady-state economics* (2d ed.). Washington, DC: Island Press.

Daly, H. E. 1993. The perils of free trade. *Scientific American* 269 (November): 50–57.

Daly, H. E., and J. B. Cobb, Jr. 1994. *For the common good.* Boston: Beacon Press.

Danaher, K. 1997. *Corporations are gonna get your momma: Globalization and the downsizing of the American dream.* Monroe, ME: Common Courage Press.

Demeny, P. 1990. Population. In *The earth as transformed by human action,* ed. B. L. Turner II, W. C. Clark, R. W. Kates, J. F. Richards, J. T. Mathews, and W. B. Meyer, 41–54. New York: Cambridge University Press.

Dobb, E. 1996. Pennies from Hell: In Montana, the bill for America's copper comes due. *Harper's Magazine* 293 (October): 39–54.

Dunning, J. H. 1999. *Governments, globalization, and international business.* New York: Oxford.

Economist. 1997a. Plenty of gloom. *The Economist* 395 (December 20): 19–21.

Economist. 1997b. Schools brief: Globalisation. *The Economist* 345 (October 18): 79–80.

Economist. 1998. An invaluable environment. *The Economist* 346 (April 18): 75.

Economist. 1999a. Why greens should love trade. *The Economist* 353 (October 9): 17–18.

Economist. 1999b. Embracing greenery. *The Economist* 353 (October 9): 89–90.

ESRI. 1995. *Understanding GIS: The ARC/INFO method*. Technical report, Environmental Systems Research Institute, Redlands, CA.

Febvre, L. 1924. *A geographical introduction to history*. New York: Barnes & Noble.

Forrester, J. W. 1961. *Industrial dynamics*. Cambridge, MA: Productivity Press.

Forrester, J. W. 1969. *Urban dynamics*. Cambridge, MA: Productivity Press.

Forrester, J. W. 1971. Counterintuitive behavior of social systems. *Technology review* 73 (January): 52–68.

Forrester, J. W. 1973. *World dynamics* (2d ed.) Cambridge, MA: MIT Press.

Forrester, J. W. 1989. *The system dynamics national model: Macrobehavior from microstructure*. Technical report D-4020, MIT System Dynamics Group, Cambridge, MA.

Friedman, T. L. 1999. *The lexus and the olive tree*. New York: Farrar, Straus, and Giroux.

Fukuyama, F. 1989. The end of history? *The National Interest* 18: 3–18.

Fukuyama, F. 1992. *The end of history and the last man*. New York: Free Press.

Fuller, T. 1987. Jeremy Bentham and James Mill. In *History of political philosophy* (3d ed.), ed. L. Strauss and J. Cropsey, 710–731. Chicago: University of Chicago Press.

Gaylin, W. 1989. Interview with Willard Gaylin. In *A world of ideas*, ed. B. Moyers, 119–126. New York: Doubleday.

Gereffi, G. 1983. *The pharmaceutical industry and dependency in the third world*. Princeton, NJ: Princeton University Press.

Getis, A., and J. Ord. 1992. The analysis of spatial association by use of distance statistics. *Geographical Analysis* 24: 189–206.

Gleick, J. 1987. *Chaos: Making a new science*. New York: Viking.

Goldsmith, E., P. Bunyard, N. Hildyard, and P. McCully. 1990. *Imperiled planet*. Cambridge, MA: MIT Press.

Goodman, M. R. 1974. *Study notes in system dynamics*. Cambridge, MA: MIT Press.

Graham, W. 1996. Masters of the game: How the U.S. protects the traffic in cheap Mexican labor. *Harper's Magazine* 293 (July): 35–50.

Gray, J. 1999. *False dawn: The delusions of global capitalism*. New York: New Press.

Greider, W. 1997. *One world, ready or not: The manic logic of global capitalism*. New York: Simon and Schuster.

Grossman, G. M., and A. B. Krueger. 1993. Environmental impacts of a North American free trade agreement. In *The Mexico-U.S. free trade agreement*, ed. P. M. Garber, 13–56. Cambridge, MA: MIT Press.

Haas, P. M., R. O. Keohane, and M. A. Levy. 1993. *Institutions for the earth: Sources of international environmental protection*. Cambridge, MA: MIT Press.

Haas, P. M., and J. Sundgren. 1993. Evolving international environmental law: Changing practices of international sovereignty. In *Global accord: Environmental challenges and international responses*, ed. N. Choucri, 401–429. Cambridge, MA: MIT Press.

Hall, N. 1991. *Exploring chaos: A guide to the new science of disorder*. New York: W. W. Norton.

Halpern, S. L. 1992. *United Nations conference on environment and development: Process and documentation*. Providence, RI: Academic Council for the United Nations System.

Hardin, G. 1968. The tragedy of the commons. *Science* 162: 1243–1248.

Hart, H. 1961. *The concept of law*. New York: Oxford.

Headrick, D. R. 1990. Technological change. In *The earth as transformed by human action*, ed. B. L. Turner II, W. C. Clark, R. W. Kates, J. F. Richards, J. T. Mathews, and W. B. Meyer, 55–67. New York: Cambridge University Press.

Hirst, P., and G. Thompson. 1999. *Globalization in question*. New York: Polity.

Hofstadter, D. R. 1979. *Gödel, Escher, Bach*. New York: Vintage Books.

Homer-Dixon, T. F. 1991. On the threshold: Environmental changes as causes of acute conflict. *International Security* 16: 76–116.

Homer-Dixon, T. F. 1993. Physical dimensions of global change. In *Global accord: Environmental challenges and international responses*, ed. N. Choucri, 43–66. Cambridge, MA: MIT Press.

Homer-Dixon, T. F. 1994. Environmental scarcities and violent conflict: Evidence from cases. *International security* 19: 5–40.

Homer-Dixon, T. F. 1999. *Environment, scarcity, and violence*. Princeton, NJ: Princeton.

Homer-Dixon, T. F. 2000. *The ingenuity gap*. New York: Knopf.

Homer-Dixon, T. F., J. H. Boutwell, and G. W. Rathjens 1993. Environmental change and violent conflict. *Scientific American* 268: 38–45.

Hoogvelt, A., and L. S. Popova. 1997. *Globalization and the postcolonial world: The new political economy of development*. Baltimore, MD: Johns Hopkins University Press.

Houghton, R. A., and D. L. Skole. 1990. Carbon. In *The earth as transformed by human action*, ed. B. L. Turner II, W. C. Clark, R. W. Kates, J. F. Richards, J. T. Mathews, and W. B. Meyer, 393–480. New York: Cambridge University Press.

HPS. 1990. Stella II User's Guide. Technical report, High Performance Systems, Hanover, NH.

HPS. 1997. An introduction to systems thinking. Technical report, High Performance Systems, Hanover, NH.

Hurrell, A., and B. Kingsbury. 1992. *The international politics of the environment.* Oxford, UK: Clarendon Press.

International Monetary Fund. 1994. *Direction of trade.* Ann Arbor, MI: Inter-university Consortium for Political and Social Research.

Isaak, R. A. 2000. *Managing world economic change: International political economy* (3d ed.). Saddle River, NJ: Prentice Hall.

Jaggers, K., and T. R. Gurr. 1995. Tracking democracy's third wave with the Polity III data. *Journal of peace research* 32(4): 469–482.

Jameson, F., and M. Miyoshi. 1998. *The cultures of globalization.* Durham, NC: Duke University Press.

Kaplan, R. D. 1993. *Balkan ghosts: A journey through history.* New York: St. Martin's Press.

Kaplan, R. D. 1994. The coming anarchy. *Atlantic Monthly* 273: 44–76.

Karliner, J. 1997. *The corporate planet: Ecology and politics in the age of globalization.* San Francisco: Sierra Club Books.

Kates, R. W., B. L. Turner II, and W. C. Clark. 1990. The great transformation. In *The earth as transformed by human action,* ed. B. L. Turner II, W. C. Clark, R. W. Kates, J. F. Richards, J. T. Mathews, and W. B. Meyer, 1–17. New York: Cambridge University Press.

Kegley Jr., C. W., and E. R. Wittkopf. 1995. *World politics: Trend and transformation* (5th ed.). New York: St. Martin's.

Kennedy, D. M. 1996. Can we still afford to be a nation of immigrants? *Atlantic Monthly* 278 (November): 52–68.

Kennedy, P. 1992. *A guide to econometrics* (3d ed.). Cambridge, MA: MIT Press.

Keohane, R. O. 1986a. *Neorealism and its critics.* New York: Columbia University Press.

Keohane, R. O. 1986b. Realism, neorealism, and the study of politics. In *Neorealism and its critics,* ed. R. O. Keohane, 1–26. New York: Columbia University Press.

Keohane, R. O. 1986c. Theory of world politics: Structural realism and beyond. In *Neorealism and its critics,* ed. R. O. Keohane, 158–203. New York: Columbia University Press.

Kindleberger, C. P. 1962. *Foreign trade and the national economy.* New Haven, Conn.: Yale.

Kindleberger, C. P. 1973. *The world depression, 1929–1939.* New York: Penguin.

Korten, D. 1996. *When corporations rule the world.* Bloomfield, CT: Kumarian.

Korten, D. 1998. *Globalizing civil society: Reclaiming our rights to power.* New York: Seven Stories.

Krasner, S. D. 1982. Structural causes and regime consequences: Regimes as intervening variables. *International Organization* 36: 185–205.

Krasner, S. D. 1983. *International regimes*. Ithaca, NY: Cornell University Press.

Kristof, N. D., and D. E. Sanger. 1999. How U.S. wooed Asia to let cash flow in. *New York Times* (February 16). Global Contagion Series, Part 2.

Kristof, N. D., and S. WuDunn. 1999a. The world's ills may be obvious, but their cure is not. *New York Times* (February 18). Global Contagion Series, Part 4.

Kristof, N. D., and S. WuDunn. 1999b. World's markets, none of them an island. *New York Times* (February 17). Global Contagion Series, Part 3.

Kristof, N. D., and E. Wyatt. 1999. Who sank, or swam, in choppy currents of a world cash ocean. *New York Times* (February 15). Global Contagion Series, Part 1.

Krugman, P. 1991. *Geography and trade*. Cambridge, MA: MIT Press.

Kuttner, R. 1999. *Everything for sale*. Chicago: Chicago.

Lane, P. 1998a. Turtle wars. *The Economist* 349 (October 3): 22–24. Survey: World Trade, Part 7.

Lane, P. 1998b. Why trade is good for you. *The Economist* 349 (October 3): 4–6. Survey: World Trade, Part 2.

Lechner, F. J., and J. Boli 1999. *The globalization reader*. New York: Blackwell.

Leithold, L. 1976. *The calculus with analytic geometry* (3d ed.). New York: Harper and Row.

Litvin, D. 1998a. Loaves and fishes. *The Economist*. Survey: Development and the Environment, Part 6.

Litvin, D. 1998b. Stumped by trees. *The Economist*. Survey: Development and the Environment, Part 5.

Lofdahl, C. L. 1992. Relationships between aggregate human activity, the environment, and global politics: A computer simulation. Master's thesis, Massachusetts Institute of Technology, Cambridge, MA.

Maddala, G. 1992. *Introduction to econometrics* (2d ed.). New York: Macmillan.

Malthus, T. R. 1976 [1798]. *An essay on the principle of population*. New York: W. W. Norton.

Mandelbrot, B. 1991. Fractals—A geometry of nature. In *Exploring chaos: A guide to the new science of disorder,* ed. N. Hall, 122–135. New York: W. W. Norton.

Mander, J., and E. Goldsmith. 1996. *The case against the global economy*. San Francisco: Sierra Club Books.

Margolis, H. 1987. *Patterns, thinking, and cognition: A theory of judgment*. Chicago: University of Chicago Press.

Marland, G., T. A. Boden, R. C. Griffin, S. F. Huang, P. Kanciruk, and T. R. Nelson. 1989. Estimates of CO_2 emissions from fossil fuel burning and cement manufacturing, based on the United Nations energy statistics and the U.S. bureau of mines cement manufacturing data. Technical Report ORNL/CDIAC-25, NDP-030, Carbon Dioxide Information Analysis Center, Oak Ridge National Laboratory, Oak Ridge, Tenn.

Marquardt, M. J. 1998. *The global advantage: How world class organizations improve performance through globalization.* Houston, TX: Gulf Publishing.

Maser, C. 1988. *The redesigned forest.* San Pedro, CA: R.& E. Mills.

Mathews, J. T. 1989. Redefining security. *Foreign Affairs* 46(2): 162–177.

Mathews, J. T. 1992. Coping with the uncertainties of the greenhouse effect. In *The global agenda* (3d ed.), eds. C. W. Kegley Jr., and E. R. Wittkopf, 366–372. New York: McGraw-Hill.

McC. Adams, R. 1990. Foreward: The relativity of time and transformation. In *The earth as transformed by human action,* eds. B. L. Turner II, W. C. Clark, R. W. Kates, J. F. Richards, J. T. Mathews, and W. B. Meyer, vii–x. New York: Cambridge University Press.

McCloskey, D. N. 1995. Computation outstrips analysis. *Scientific American* 273(1): 26.

Meadows, D. H., D. L. Meadows, and J. Randers. 1992. *Beyond the limits.* Post Mill, VT: Chelsea Green Publishing.

Meadows, D. H., D. L. Meadows, J. Randers, and W. W. Behrens III. 1972. *The limits to growth.* New York: Universe Books.

Midlarsky, M. I. 1989. *Handbook of war studies.* Ann Arbor: University of Michigan Press.

Mitchell, R. 1993. Intentional oil pollution of the oceans. In *Institutions for the earth: Sources of international environmental protection,* ed. P. M. Haas, R. O. Keohane, and M. A. Levy, 183–247. Cambridge, MA: MIT Press.

Mitchell, R. B. 1994. *International oil pollution at sea: Environmental policy and treaty compliance.* Cambridge, MA: MIT Press.

Morecroft, J. D. 1983. System dynamics: Portraying bounded rationality. *International journal of management science* 11(2): 131–142.

Morgenthau, H. J. 1985 [1948]. *Politics among nations: The struggle for power and peace* (6th rev. ed.). New York: A. A. Knopf.

Most, B., and H. Starr. 1989. *Inquiry, logic, and international politics.* Columbia: University of South Carolina Press.

Nicolis, G., and I. Prigogine. 1989. *Exploring complexity.* New York: W. H. Freeman.

Norris, F. 2001. The U.S. trade deficit now matters. *New York Times,* (February 23).

North, R. C. 1990. *War, peace, survival: Global politics and conceptual synthesis.* Boulder, CO: Westview Press.

North, R. C., and N. Choucri. 1996. New perspectives on the fourth image: Report on a global system. Presented at the International Studies Association annual convention, San Diego, CA, April 16–20. (Cited with permission.)

Ohmae, K. 1999. *The borderless world: Power and strategy in the interlinked economy.* New York: Harper Business.

O'Loughlin, J., and L. Anselin. 1992. Geography of international conflict and cooperation: Theory and methods. In *The new geopolitics,* ed. M. D. Ward, 11–38. Philadelphia: Gordon and Breach.

Olson, J. S., J. A. Watts, and L. J. Allison. 1985. Major world ecosystem complexes ranked by carbon in live vegetation: A database. Technical Report NDP-017, Carbon Dioxide Information Center, Oak Ridge National Laboratory, Oak Ridge, Tenn.

Ophuls, W. 1977. *Ecology and the politics of scarcity.* San Francisco: W. H. Freeman.

Ord, J., and A. Getis. 1995. Local spatial autocorrelation statistics: Distributional issues and an application. *Geographical analysis* 27(4): 286–305.

Ostrom, E. 1990. *Governing the commons: The evolution of institutions for collective action.* New York: Cambridge University Press.

O'Tuathail, G. 1996. *Critical geopolitics: The politics of writing global space.* Minneapolis: University of Minnesota Press.

Panayotou, T. 1992. The economics of environmental degradation: Problems, causes, and responses. In *Environmental economics: A reader,* ed. A. Markandya and J. Richardson, 316–363. New York: St. Martin's Press.

Pearce, D. W. 1992. *The MIT dictionary of modern economics* (4th ed.). Cambridge, MA: MIT Press.

Penubarti, M. 1994. Market openness and U.S.–Japan trade conflict. Ph.D. dissertation, University of Colorado, Boulder, CO.

Plato. 1991. *The republic,* (trans. Allan Bloom 2d ed.). New York: Basic Books.

Ponting, C. 1991. *A green history of the world: The environment and the collapse of great civilizations.* New York: St. Martin's Press.

Prigogine, I. 1980. *From being to becoming: Time and complexity in the physical sciences.* San Francisco: W. H. Freeman.

Prigogine, I., and I. Stengers 1984. *Order out of chaos: Man's new dialogue with nature.* New York: Bantam Books.

Putnam, R. D. 1988. Diplomacy and domestic politics: The logic of two-level games. *International Organization* 42: 427–460.

Rand McNally. 1982. *Illustrated atlas of the world.* Chicago: Rand McNally.

Randers, J. 1980. Guidelines for model conceptualization. In *Elements of the system dynamics method,* ed. J. Randers, 117–139. Cambridge, MA: Productivity Press.

Reddy, R. 1996. The challenge of artificial intelligence. *Computer* 29: 86–98.

Renouvin, P., and J. B. Duroselle. 1964. *Introduction to the history of international relations.* New York: Praeger.

Ricardo, D. 1911 [1817]. *The principles of political economy and taxation.* New York: E. P. Dutton.

Richards, J. F. 1990. Land transformation. In *The earth as transformed by human action*, ed. B. L. Turner II, W. C. Clark, R. W. Kates, J. F. Richards, J. T. Mathews, and W. B. Meyer, 163–178. New York: Cambridge University Press.

Richardson, G. P. 1991. *Feedback thought in social science and systems theory.* Philadelphia: University of Pennsylvania.

Richardson, J. M. 1983. The usefulness of global models. Technical report D-3409, MIT System Dynamics Education Project, Cambridge, MA.

Roberts, C., and D. Roberts. 1980. *A history of England: 1688 to present.* Englewood Cliffs, NJ: Prentice-Hall.

Rodrik, D. 1997. *Has globalization gone too far?* Washington, DC: Institute for International Economics.

Rodrik, D. 1999. *The new global economy and developing countries: Making openness work.* Baltimore, MD: Johns Hopkins.

Rowell, J. 1995. Supercharged computers: powerful tool for solving challenges of science. The Institute for Science, Engineering and Public Policy presents Dr. Stephen W. Hawking (July 11).

Rucker, R. 1987. *Mindtools: The five levels of mathematical reality.* Boston: Houghton Mifflin.

Ruggie, J. G. 1986. Continuity and transformation in the world polity: Toward a neorealist synthesis. In *Neorealism and its critics*, ed. R. O. Keohane, 131–157. New York: Columbia University Press.

Sagan, C. 1980. *Cosmos.* New York: Random House.

Samuelson, P. A., and W. D. Nordhaus. 1992. *Economics* (14th ed.). New York: McGraw-Hill.

Saurin, J. 1993. Global environmental degradation, modernity and environmental knowledge. *Environmental politics* 2(4): 46–64.

Schneider, S. H., and P. J. Boston. 1991. *Scientists on gaia.* Cambridge, MA: MIT Press.

Schol, E. 1995. BluediveII.mpeg. computer graphics file, University of Twente, Department of Computer Science, Enschede, The Netherlands. A descent into a blue Mandelbrot landscape. Available at <http//www.student.utwente.nl/~schol/gallery/>.

Shiva, V. 1997. *Biopiracy.* Cambridge, MA: South End Press.

Shiva, V. 1999. *Stolen harvest: The hijacking of the global food supply.* Cambridge, MA: South End Press.

Simon, H. A. 1983. *Reason in human affairs.* Stanford, CA: Stanford University Press.

Simon, H. A. 1985. Human nature in politics. *American Political Science Review* 79: 293–304.

Smith, A. 1869 [1759]. Theory of moral sentiments. In *The essays of Adam Smith.* London: Alexander Murray.

Smith, A. 1937 [1776]. *An inquiry into the nature and causes of the wealth of nations.* New York: Modern Library.

Soroos, M. S. 1994. Confronting global environmental change. *Mershon International Studies Review* 38(2): 299–306.

Sprout, H., and M. Sprout. 1965. *The ecological perspective on human affairs—With special reference to international politics.* Princeton, NJ: Princeton University Press.

Sprout, H., and M. Sprout. 1968. *An ecological paradigm for the study of international politics.* Princeton, NJ: Center of International Studies, Princeton University.

Sprout, H., and M. Sprout. 1971. *Toward a politics of the planet earth.* New York: Van Norstrand Reinhold.

Starr, H. 1992. Joining political and geographic perspectives: Geopolitics and international relations. In *The new geopolitics,* ed. M. D. Ward, 1–9. Philadelphia: Gordon and Breach.

StatSci. 1995. *S-Plus guide to statistical and mathematical analysis* (3.3 ed.). Seattle, WA: StatSci. a division of MathSoft, Inc.

Sterman, J. D. 2000. *Business dynamics: Systems thinking and modeling for a complex world.* New York: McGraw Hill.

Stewart, I. 1989. *Does God play dice?: The mathematics of chaos.* New York: Basil Blackwell.

Strong, M. F. 1993. In *Global accord: Environmental challenges and international responses,* ed. N. Choucri, ix–xiii. Cambridge, MA: MIT Press.

Summers, R., and A. Heston. 1991. The Penn world table (mark 5): An expanded set of international comparisons, 1950–1988. *Quarterly Journal of Economics* 106(9): 1–41.

Sykes, C. 1993. *The best mind since Einstein?* Boston: WGBH. An hour-long television biography on physicist Richard Feynman, produced for the PBS series *NOVA* by BBC-TV.

Taylor, P. J. 1993. *Political geography: World–economy, nation–state, and locality* (3d ed.). New York: John Wiley and Sons.

Thompson, J., and H. Stewart. 1986. *Nonlinear dynamics and chaos.* New York: John Wiley and Sons.

Thucydides. 1972. *History of the Peloponnesian war.* New York: Penguin Books.

Tolba, M. K., and O. A. El-Kholy. 1992. *The world environment, 1972–1992.* New York: Chapman and Hall.

Turner, P. 1996. The politics of tropical deforestation: Implications for the second image. Presented at the International Studies Association annual convention, San Diego, CA, April 16–20.

Turner II, B. L., W. C. Clark, R. W. Kates, J. F. Richards, J. T. Mathews, and W. B. Meyer. 1990. *The earth as transformed by human action: Global and regional changes in the biosphere over the past 300 years*. New York: Cambridge University Press.

Ullman, R. 1983. Redefining security. *International Security* 8: 129–153.

UNFAO 1993. Forest resources assessment 1990: Tropical countries. Technical report, FAO Forestry Paper 112, Food and Agriculture Organization of the United Nations, Rome.

UNFAO 1995. Forest resources assessment 1990: Global synthesis. Technical report, FAO Forestry Paper 124, Food and Agriculture Organization of the United Nations, Rome.

Wallerstein, I. 1979. *The capitalist world-economy*. New York: Cambridge University Press.

Walt, S. M. 1991. The renaissance of security studies. *International Studies Quarterly* 16(2): 221–239.

Waltz, K. N. 1959. *Man, the state and war: A theoretical analysis*. New York: Columbia University Press.

Waltz, K. N. 1979. *Theory of international politics*. Reading, MA: Addison-Wesley.

Waltz, K. N. 1986. Reflections on Theory of International Politics: A response to my critics. In *Neorealism and its critics,* ed. R. O. Keohane, 322–345. New York: Columbia University Press.

Ward, M. D. 1992a. *The new geopolitics*. Philadelphia: Gordon and Breach.

Ward, M. D. 1992b. Throwing the state back out. In *The new geopolitics,* ed. M. D. Ward, vii–x. Philadelphia: Gordon and Breach.

Ward, M. D., D. R. Davis, and C. L. Lofdahl 1995. A century of tradeoffs: Defense and growth in Japan and the United States. *International Studies Quarterly* 39(1): 27–50.

Ward, M. D., and C. L. Lofdahl. 1995. Toward a political economy of scale: European integration and disintegration. In *Towards a new Europe: Stops and starts in regional integration,* ed. G. Schneider, P. A. Weitsman, and T. Bernauer, 11–27. Westport, CT: Praeger.

Weiss, E. B. 1992. *Environmental change and international law: New challenges and dimensions*. Tokyo: United Nations University Press.

Whitmore, T. M., B. L. Turner II, D. L. Johnoson, R. W. Kates, and T. R. Gottschang. 1990. Long-term population change. In *The earth as transformed by human action,* ed. B. L. Turner II, W. C. Clark, R. W. Kates, J. F. Richards, J. T. Mathews, and W. B. Meyer, 25–39. New York: Cambridge University Press.

Williams, M. 1990. Forests. In *The earth as transformed by human action,* ed. B. L. Turner II, W. C. Clark, R. W. Kates, J. F. Richards, J. T. Mathews, and W. B. Meyer, 179–201. New York: Cambridge University Press.

Williams, M. 1993. International trade and the environment: Issues, perspectives and challenges. *Environmental Politics* 2(4): 80–97.

Wilmsen, S. K. 1991. *Silverado: Neil Bush and the savings and loan scandal.* Washington, DC: National Press Books.

Wils, A., M. Kamiya, and N. Choucri. 1998. Threats to sustainability: Simulating conflict within and between nations. *System Dynamics Review* 14(2–3): 129–162.

Wilson, E. O. 1998. *Consilience: The unity of knowledge.* New York: Knopf.

Woo, W. T. 1990. The art of economic development: Markets, politics, and externalities. *International Organization* 44(3): 403–429.

World Bank. 1995. *Social indicators of development.* Washington, DC: Socio-Economic Data Division, International Economics Dept., World Bank.

World Resources Institute. 1996. *World resources 1996–1977: A guide to the global environment.* New York: Basic Books.

Young, O. R. 1989. *International cooperation: Building regimes for natural resources and the environment.* Ithaca, NY: Cornell University Press.

Young, O. R. 1993. Negotiating an international climate regime: The institutional bargaining for environmental governance. In *Global accord: Environmental challenges and international responses,* ed. N. Choucri, 431–452. Cambridge, MA: MIT Press.

Index

Absolutism, 21
Abstraction, 37, 42, 46, 61, 137
Academy, natural and social split within, 222n17
Activists, anti-trade, 1–2, 155, 167
Adaptation, 6
Advantage, comparative. *See* Comparative advantage
Afforestation, 76, 91–92, 123, 204
Africa, 7, 30, 58, 70, 108–110, 121–122, 190
Aggregation, 20, 22–23, 28, 37–40, 42, 69, 78, 81–82, 97, 135–137, 151, 162, 167
Agnew, J. A., 221n3
Agricultural expansion, 86–88, 94–96, 114, 186–187, 225n41
Agricultural politics, 11–12
Agricultural production, 131
Agricultural trade, 3, 8, 85, 96
Akira, S., 190
Algebraic curves, 173
Alt, J. E., 51–52, 215
America, 6–9, 33
Amin, S., 8
Anarchy, 35–36, 40
Anglo-American, international economy dominance of, 9, 13–14
Anthropocentric environmental change, 4, 160
Antiglobalization. *See* Seattle riots

Antitrade. *See* Seattle riots
Anti-WTO. *See* Seattle riots
Arabia. *See* Saudi Arabia
Arable land, 128
Archimedes, spiral of, 173–174
Aron, R., 17–18
Ashley, R. K., 221
Asia, 8, 30, 108–109, 189–190
Asymmetric trade relationships, 60, 152, 157, 166
Athens, 61–62
Atmospheric effluents. *See* Carbon dioxide (CO_2)
Attractor, chaotic, 173–174
Australia, 70–71, 106, 179–180, 223–224n28
Autocorrelation, 55, 120, 122, 191
 spatial, 196–199
 temporal, 195–196
Axelrod, R., 36, 163

Baker, D., 5, 9
Balancing causal relationships. *See* Negative feedback
Balkans, 71
Barnet, R. J., 8
Beck, N., 196–197
Behavioral consequences, 38
Behavioral modeling, 21–28, 40
Behavioral regularity, 32–33
Bennett, J. W., 92, 94
Berlin Wall, 65, 109

Bhagwati, J., 66, 100–104, 106, 115–116, 124–125, 225n36
Biogeochemical cycles, 74, 224n29
Biological diversity, 30
Biological metaphor, 17–18
Biomass, main components of, 224n29
Biome, 76–77
Biosphere, 96
Biotic-carbon density map, 76–77
Birth rates, 91, 202
  ELP definition of, 131–134, 137, 142, 203
Bishop, C. M., 42
Bivariate statistical analysis, 111–116
Boulding, K. E., 221
Bounded rationality, 20, 25–26, 39, 42, 47, 57–58, 222n10
Bradley, E., 173–176
Braudel, F., 21, 190
Brazil, 66, 70, 120, 168, 223–224n28
Brecher, J., 8
Bretton Woods, 13, 158
Brzezinski, Z., 14, 47
Bulgaria, 70, 109, 189, 223–224n28
Burtless, G., 6

CO_2. *See* Carbon dioxide
Calculus, 173
  expected utility, 21, 37
  power, 17
Canada, 70, 108, 179–183, 199, 223–224n28
Capitalism, 9–16, 65
  criticism of excesses, 96
Carbon dioxide (CO_2), 53, 72–99, 109–116, 160–162, 180–181, 222n15, 223n25, 224n33, 225n40
  CO_2 per capita, 72–85, 106–114, 124–125, 181, 189–190
Carr, E. H., 13, 36
Cassidy, J., 16, 22
Causal-loop diagrams, 59, 133, 141–144, 150

Causation, 19, 23–27, 34, 55, 60–61, 67, 78, 96–97, 103, 112–114, 123–124, 133–136, 141–145, 155, 159–165, 219–220, 221n8, 222n15
Cavanagh, J., 8
Centralization, 5, 13, 33, 37, 40, 47–48, 61–62
CFC (chlorofluorocarbons), 222n15
Challenge, hegemonic, 14
Chaos theory, 132, 171–176
China, 70, 120, 177, 180, 181, 183, 184, 223–224n28
Chiras, D. D., 224n29
Chisholm, M., 14, 51, 82, 85, 97
Choucri, N., 30, 34, 40–44, 48–57, 68–74, 79–81, 86, 89, 92, 96, 137, 159–160, 167, 180–181, 222n14, 223nn24, 27, 224n31, 226n51
Classical economics. *See* Economics
Classification, 42
Cliff, A., 55, 198
Climate change, 53
Clinton, W. J., 169
Coal, 82
Coca-Cola, 6
Coincident processes, 30, 66–67, 103, 114, 149
Cold War, 6, 14, 18–20, 61–62, 65–66, 99
Cole, S., 135
Colgate-Palmolive, 6
Collapse, Soviet Union/Communism, 14, 62, 65
Colonialism, 7, 14, 86
Colonization, 79
Combustion engine, 85, 94
Commerce, expanding international, 102
Commodities, 85, 135, 148, 154, 215
Common Pool Resources (CPRs), 47–48
Commons, tragedy of the. *See* Tragedy of the commons
Communications technology, advanced, 5, 8, 96, 117, 169

Companies. *See* Multi-National Corporations (MNC)
Comparative advantage, 3, 12, 50–51, 100, 102, 127–130, 136, 140, 144, 147, 149, 153, 156, 207, 209, 215–216
Complexity, 9–10, 15–16, 19–28, 37–40, 50–52, 58, 77, 82, 85–86, 96, 99, 103–105, 108, 124–126, 151, 158, 161–170, 211
Computation, advances in, 10, 22–27, 105, 127, 130–131, 158, 161–164, 171–176
Conflict, 17–19, 31, 35, 40–41, 47, 52, 62, 68, 168, 223n24. *See also* Cooperation
Consequences, unintended. *See* Unintended consequences
Consumerism, 6
Consumption, 12, 50–52, 82–87, 102, 129–134, 156, 161
 ELP definition of, 137–143, 151–152, 204
Contagion effect, 86, 169, 223n21. *See also* Diffusion
Context, definition of, 21
Contiguity matrices, 55, 107–108, 111, 117–118, 162, 199–200, 225–226n43
Cooperation, 17, 33, 35. *See also* Conflict
Core, 21–22, 51. *See also* Periphery
Cornell University, 221
Corporations. *See* Multi-National Corporations (MNCs)
Correlation, 28, 54, 67, 73, 78–79, 96–97, 99, 103–105, 111–114, 124, 152, 162–163, 166, 194, 225–226n43
 spatial, 109–112, 120–121
Cosmos, 31–32, 61
Cost benefit analysis, 21, 48
Costs, 99–105, 124–126, 155–157
 economic, 7, 15, 44, 50, 52, 85, 169–170, 213

 environmental, 10, 16, 48, 51–52, 82, 127–128, 136, 148, 150, 153, 166, 219, 225n36
Counterfactual analysis, 43
Counterintuitive policy consequences, 165–167
CPRs. *See* Common Pool Resources
Cropsey, J., 11–12, 39, 63, 96
Cross-national analysis, 56, 69, 92, 105, 108, 111, 126, 177, 197
Cross-sectional statistical test, 68, 77, 191, 195, 197–200
Cuba, 71

Daly, H., 46, 100–106, 124–125, 224n32, 225n40
Danaher, K., 8
Davos. *See* World Economic Forum
DDT (dichloro-diphenyl-trichloro-ethane), 94
Death rates, 91
 ELP definition of, 132–134, 137, 142, 203
Debate, trade and the environment, 4–5, 9, 27, 99–106, 125, 128, 130–131, 153, 165–166
Debt, foreign, 30, 66
Decentralization, 13, 33
Decolonization, 14
Deductive analysis, 22, 35–36, 56
Definitions, model. *See* Model definitions
Deforestation, 30, 53, 67, 74–78, 88, 91–96, 99, 106–107, 110–112, 115–116, 121–124, 127, 152, 157, 162–163, 168, 185–186, 189–190, 199, 222nn15, 18, 223n25, 225n42
Delegates, trade, 1, 155, 167
Demeny, P., 89–91, 223n24
Democracy, 6, 9
Democritus, 32
Depreciation, ELP definition of, 137, 142–143, 203
Depression, economic, 2, 8, 13, 158
Desert, biome definition, 77
Determinism, 17, 20

Development, economic, 8–13, 15, 17, 21, 42, 46, 65–67, 73, 82, 84, 116, 124, 152, 156–157, 166–167
dFW (delta Forest and Woodland area, i.e., Forest Change), 191–194
Diagnostic analysis, 191–195
Differential equations, 23–26, 132, 138, 159, 164
model definitions, 201–210
Diffusion, 54–56, 61, 161–163, 223n21. *See also* Contagion effect
model definitions, 117–121
Dimensionality, systemic representation of, 22, 37, 92, 96, 105
Disputes, trade, 2, 66, 101, 221n1, 223n26
Diversity
biological, 30
international standards, 104, 215
national endowments, 130
Dobb, E., 101
Dolphin. *See* Tuna and dolphin trade dispute
Domestic economics, 2, 40, 50, 87, 91, 102–104
Domestic environments, 9, 86, 94–96, 101, 106, 152
Domestic politics, 38, 43–44, 115–117
Domestic variables, 121–124, 186
Duffing oscillator, 175
Dynamic equilibrium, 219–220

East Asia, 7, 169
Eastern Europe, 17–20, 71, 180, 189, 224n33
Economist, The, 4–5, 15, 135, 154, 156, 165
Economics, 2, 5, 8, 10–12, 15–16, 21–23, 25–31, 49–52, 67, 80, 100–102, 128, 158–159, 167–168
Economy, global, 6, 10–12, 15, 21, 49–50, 58, 82, 94, 102, 104, 116, 130, 135, 154, 157–158, 161, 166, 190
Ecosystems, 29–30, 94

Efficiency, 94, 142, 168, 208, 214
economic, 3, 38, 50–51, 101–104, 125, 130
energy, 83–85
ELP (Environmental Lateral Pressure), 58–61, 149–154, 226n48
model definition, 202–209
model development, 136–149
Emergence, 36
Emigration. *See* Migration
Emissions. *See* Carbon dioxide (CO_2)
Emotionalism, environmental, 1–2, 100, 103, 124, 221n7
Empedocles, 32
Empirical analysis, outline of, 58–61
Endogenaity, 26, 216
Endowments, national, 17, 32, 38, 42, 69, 77, 130, 153
England. *See* Great Britain
Entity-environment relationship, 20–22, 222n12
Entropy, 82
Environments, natural and social, 10–22, 26–27, 30–31, 34, 45, 49, 52–54, 58, 67–68, 74–75, 78, 99, 105, 128, 131, 142–143, 152, 159–163, 171, 176
Environmental Lateral Pressure. *See* ELP
Environmental costs. *See* Costs, environmental
Environmentalists, 2–5, 15, 27, 32–33, 99–105, 115, 124–125, 155–156, 166
Equilibrium, dynamic. *See* Dynamic equilibrium
ESRI (Environmental Systems Research Institute), 54, 68
Europe, 8, 11, 13, 43, 70, 91, 108–110, 130, 190, 226n55. *See also* Eastern and Western Europe
Exogenaity, 26. *See also* Endogenaity
Expansion, 11, 40–44, 223n21
agricultural, 87–88, 94, 187
market economy, 5, 9, 21, 50–53, 79, 85, 143

population, 128
national, 17, 49–53, 56, 63, 86, 119
social environmental, 49, 53, 67, 75, 82–88, 160, 168
Expectations, 48, 170, 213, 222n20
Expenditures, military, 43–44
Exploitation, environmental, 7, 19
Exponential growth, 131, 134
Exports, 51, 86, 94, 118, 157, 181–183, 225–226n43
Externalities, environmental, 82, 104, 137, 157
Extraction, resource, 19, 58, 92, 204

F-statistic (model fit), 191, 195, 198, 200
Factor specificity, 51, 205, 215–216
FAO. *See* UNFAO
FDI (foreign direct investment), 5
Febvre, L., 20
Feedback, 56, 60–61, 137, 168, 175–176, 219, 221n5, 223n22
 consequences of, 163–166
 definition of, 23–27
Feedback analysis, 131–134, 141–145, 149–152
Fertilizers, overuse of, 94
Feynman, R., 221n6
Fishery management, 33
Food per capita, 133–134, 202, 211
Foresight, possibility of, 164
Forest change (dependent variable), 53, 67, 87, 91–95, 114–116, 121–125, 157, 191–196, 199–200, 222n18, 223n25, 225n41
Forestation. *See* Afforestation
Forrester, J. W., 23, 25–26, 165–166, 175, 226n45
Fractals, 171–173, 221n4
France, 70, 177, 181–183, 223–224n28
Free-market, 5, 8–9
Free-rider problem, 47–48, 51
Free trade, 11–13, 50–51, 66, 79, 100–104, 116–117, 120, 124, 128

Friedman, T. L., 6–7
Fukuyama, F. 14
Fuller, T., 226n53

GATT (General Agreement on Tariffs and Trade), 2, 13, 66, 100, 158, 223n26
Gaylin, W., 32
General Agreement on Tariffs and Trade. *See* GATT
General Linear Model. *See* GLM
Geneva, 221
Geoeconomics, 221
Geographic representation, 41, 54–55, 107, 111, 151, 162
Geography, 5, 10, 16–22, 25, 50–56, 68–77, 117, 159, 162, 222n11
Geopolitics, 10, 16–22, 77, 159, 221n3
Geopolitik, 18
Gereffi, G., 97
Germany, 12–13, 43, 70, 79, 158, 177, 181–183, 223–224n28
Getis, A., 107–108, 197
G_i statistic, 107–108, 117
$G*_i$ statistic, 108–112, 117, 120–121, 197, 225n37
Gilpin, R. G., 221
GIS (Geographic Information System), 54–55, 60, 68–77, 161–163
Gleick, J., 161, 174
GLM (General Linear Model), 225n38
Globalization, 1–10, 27, 29, 31, 46, 53, 68, 92–93, 96, 99, 104–106, 117, 124, 127, 130, 153, 155–158, 167–170
Global economy. *See* Economy, global
Global warming, 74, 222n15
GNP (Gross National Product), 53, 60, 69–80, 83, 87–90, 95–96, 99, 106, 108–109, 119–125, 127, 151–153, 156, 162–163, 177–180, 213–215, 217, 223n27, 224n32, 225nn40, 42, 226n51
 ELP definition of, 137–149, 203

GNP per capita, 78, 81–82, 84–85, 89, 99–100, 103, 106, 108–110, 112–125, 151–153, 156–157, 179–181, 189–200, 211–213, 224
  ELP definition of, 137, 139, 141–142, 146, 148, 205
Goldsmith, E., 29
Goodman, M. R., 226n47
Graham, W., 66
Graphical relationships, 211–217
Grassland, 17, 76
Gray, J., 8–9
Great Britain, 12–13, 17, 43, 70, 76, 79, 129–130, 146, 177, 181–183, 190, 223–224n28
Greenwich, 190
Greider, W., 8
Grossman, G. M., 103, 106, 115
Gross National Product. *See* GNP
Gulf War, 91, 109

Haas, P. M., 33
Hall, N., 161
Halpern, S. L., 67
Hardin, G., 47, 226n44
Hart, H. L. A., 32–33
Haushofer, K., 18
Headrick, D. R., 52, 80, 82, 85–87, 92, 94, 96
Heartland, theory of, 17–18
Heckscher-Ohlin trade model, 51, 205, 215
Hegemonic transition, 146
Heterogeneous space, 21
Heteroskedasticity, 191, 196–197
Heuristic, 57, 136
Hewlett-Packard, 6
Hirst, P., 5
Hitler, A., 18
Homer-Dixon, T. F., 62, 163, 168, 175
Hong Kong, 70, 179, 223–224n28
Hoogvelt, A., 7
Horizontal organization of space, 21–22

Houghton, R. A., 82
HPS (High Performance Systems), 24, 131, 201
Hutus, 109
Hydrogen, 224n29
Hydrography, 17
Hyper-connectedness, 23–25, 162–164

IBRD (International Bank for Reconstruction and Development), 13
Idealism, 36
Ideology, 18, 20–21
Image, 35–41, 45–46, 49, 61–63, 81–83, 159, 221n7
  first (individual), 35, 61, 81
  second (state), 35, 61
  third (international system), 35–36, 49, 61, 83
  fourth (global system), 45–46, 49, 61, 63, 83
IMF (International Monetary Fund), 3, 13, 87–88, 158
Immigration. *See* Migration
Imperialism, 38, 79, 87
Imports. *See also* Exports
Inconceivability, 223
India, 70, 177, 180, 184, 223n26, 223–224n28
Indicators, environmental, 53–55, 57, 74
Indifference curves, 105
Indigenous peoples, 94
Individualism, 7
Indonesia, 70, 169, 223–224n28
Inductive analysis, 22, 36, 68
Industrialization, 8, 14, 60, 79, 82, 94, 131, 156–157, 169, 181
Inequality, 5
Inference, 67, 78, 149
Inflation, 30
Information, 11, 20, 25, 38, 48, 54, 57, 68–69, 77, 86, 111, 140–141, 168, 173–174, 184, 201
Innovations, 74–77

Intellectuals, 12, 27, 161
Intelligence, 158–161
Interconnections, international, 43–45, 51, 105, 123, 155, 162, 169, 176
International Bank for Reconstruction and Development. *See* IBRD
International Monetary Fund. *See* IMF
International Political Economy (IPE), 5, 11, 38, 49, 52, 99
International relations, 6, 10, 16–22, 35, 39–45, 61, 65, 95
 theory development, 29–63
Investment, ELP definition of, 137, 140, 142, 146, 169, 202–203, 207–209, 213, 216
IR. *See* International relations
Isaak, R. A., 6
Isotropic plane, 22
Israel, 70, 91, 224

Jaggers, K., 14
Jameson, F., 8
Japan, 6, 17, 43–44, 70, 79, 122, 167, 169, 177–183, 223–224n28
Justice, 62

Kaplan, R. D., 62
Karliner, J., 7
Kates, R. W., 81–82, 85, 88
Katz, J. N., 197
Kegley, C. W., 13
Kennedy, D. M., 223n24
Kennedy, P., 55
Keohane, R. O., 33, 38–40, 42, 47, 57, 159, 221n9
Keynes, J. M., 13, 225n35
Kindleberger, C. P., 13, 69
Korea, 19, 70–71, 223–224n28
Korten, D., 7–8
Krasner, S. D., 33
Kristof, N. D., 169, 226n54
Krugman, P., 22
Kuttner, R., 9
Kuwait, 91, 167

Lagged variables
 spatial, 198–199
 temporal, 195, 198
Land-intensive activities, 12, 86, 94
Lane, P., 129, 223n26
Lateral pressure theory, 28–63, 72, 79, 88, 97, 119, 136–153, 156, 159–161, 164, 201–209
Latin America, 7, 30, 58, 70, 121–122
Latitudinal analysis, 77, 106–107, 120, 189–190, 225n42
Lawlessness, environmental consequences of, 62, 168
Lebensraum (living space), 18
Lechner, F. J., 5
Legal positivism, 226
Leithold, L., 173
Levers, policy, 67, 165–167
Lexus, 6
Liberal International Economic Order (LIEO), 13–15
Liberalization, trade, 1–3, 155
Limaçon, 173–174
Limits, 20, 103, 167
Limits to Growth, The, 131, 134–135, 137, 153, 226n45
Linkage challenge, 2, 5, 30, 34, 40, 45, 49, 52–55, 60–61, 74, 99, 103, 114, 152, 159–160, 165, 219
Litvin, D., 154, 168
Local scale, 2, 8, 18, 20–21, 25, 28, 30, 45–48, 68, 74–75, 82, 93, 96, 101, 103, 121–122, 152, 191, 195
Logging, 92–94
London, 102, 190
Longitudinal analysis, 77, 189–190
Loops, causal, 24–27, 133–134, 141–144, 151–152
Lorenz, E., 174–175
Luxembourg, 70, 179

Mackinder, H. J., 17–18
Macrobehaviors, 23, 26–27, 141, 171–172, 176, 221n4
Macroeconomics, 23

Maddala, G., 223n22
Malaysia, 70, 183, 223n26, 223–224n28
Malthus, T. R., 127–131, 134–136, 146, 153–154
Mandelbrot, B., 171
Mander, J., 8
Map-based analysis, 54–55
Margolis, H., 42
Marland, G., 53, 73, 83–84
Marquardt, M. J., 6
Marxism, 2, 14, 96
Maser, C., 92
Mathews, J. T., 65, 74
Matrix, contiguity, 55, 107–108, 117–118, 199–200
  trade connected, 117–119, 163, 225–226n43
McC. Adams, R., 86
McCloskey, D. N., 22
Meadows, D. H., 131, 226n45
Melian dialogue, 61–62
Mercantilists, 11
Mexico, 66, 70, 101, 179, 181, 199, 223–224n28
Microeconomics, 25, 37, 40, 42, 48, 52, 57, 226n44
Microfeatures, 23, 26–27, 141, 151, 171–172, 176, 221n4
Midlarsky, M. I., 40
Migration, 5, 8, 91–92, 109, 169, 189, 223n24, 225–226n43
Military power, 14, 19–20, 32, 43–44, 222n11
Minerals, 17, 42, 135, 153–154
Mitchell, R., 33
MNCs. *See* Multi-National Corporations
Model definitions and equations, dynamic
  environmental lateral pressure (ELP), 202–209
  overshoot and collapse, 201–202
Moore, M., 165
Morality, 47–48, 128, 146, 222n16
Moran's I, 191, 198–200

Morecroft, J. D., 25, 57–58
Morgenthau, H., 36, 222n11
Mortality, 91
  ELP definition of, 137, 203, 206, 208, 211–212
Most, B., 20, 222n12
Multicausality, 22, 97
Multidimensionality. *See* Dimensionality, systematic representation of
Multi-National Corporations (MNCs), 5, 7–9, 48–50, 157
Multivariate analysis, 55, 99, 106, 116–125, 159, 162, 196, 198

N30. *See* Seattle riots
NAFTA (North American Free Trade Act), 8, 100
Nation-state, 19–20, 32–35, 38–40, 45–46, 49–50, 61–62, 77, 102
National endowments. *See* Endowments, national
National-socialism. *See* Nazism
Natural environment. *See* Environments, natural and social
Nazism, 18, 158
Negative feedback, 25, 61, 134, 142–143, 151–152
Neoclassical economics. *See* Economics
Neorealism, 39, 52, 61
New International Economic Order (NIEO), 14
New York, 102
New Zealand, 70–71, 106, 223–224n28
Nicolis, G., 219
NIEO. *See* New International Economic Order
Nigeria, 70, 167, 224
Nineteenth century, 12, 17, 130, 190
Nitrogen, 224n29
Nonaligned countries. *See* Third World

Nonlinearity, 23–27, 56, 133, 164–165, 175–176, 201, 208, 211–217
Normalization, 119, 145
Norris, F., 86, 166
North, R. C., 40–50, 53–57, 61, 63, 68–71, 79, 81, 83, 86, 89, 96, 137, 159–160, 222n14, 223n27, 224n31, 226n51
North America, 70, 108, 130
North American Free Trade Act. *See* NAFTA
North, definition of, 66, 70–74, 106–107

O'Loughlin, J., 22
Ohmae, K., 6
Oil, 33, 71, 77, 82–85, 94, 96, 109, 135, 151, 153–154, 167, 179
Olson, J. S., 76
OPEC (Organization of Petroleum Exporting Countries), 84–86, 179
Ophuls, W., 47
Opportunity and willingness framework, 20–22, 222n12
Ord, J., 55, 107–108, 197–198
Ostrom, E., 32, 45, 47–48, 51, 57
O'Tuathail, G., 221n3
Outside-in-ness, 36, 46
Overfishing, 30
Overgrazing, 30
Overshoot and collapse, 131–136, 201–202
Oxygen, 73, 224n29

Pacific Ocean, 189–190
Pakistan, 70, 223n26, 223–224n28
Panayotou, T., 30, 149
Panel Corrected Standard Errors (PCSE), 191, 196–197
Paraguay, 70, 191, 195, 223–224n28
Partial differential equations. *See* Differential equations
Pattern matching, 42
PCSE. *See* Panel Corrected Standard Errors

Pearce, D. W., 12
Peloponnesian War, 61
Penubarti, M., 51
Periphery, 21–22, 51
Perturbations, systemic, 175
Pesticides, 94
Petroleum. *See* Oil
Philosophy, 7, 31–32, 35, 45, 103, 176, 221n7, 226n53
Phosphorus, 224n29
Photography, satellite. *See* Satellite photography
Physics, 23, 32, 82
Physiocrats, 11–12
"Plague, The," 62
Plantations, 92
Plato, 31–32, 61
Polar coordinates, 173
Polarity, 60–61, 133–134
Police, 1, 4, 47, 222n16
Policy consequences, 158, 164–167
Polis, 31–32, 61
Politia, 222n16
Political economy. *See* International political economy
Polity, 222
Pollution, 9, 30, 32, 51, 103, 131
Ponting, C., 30
Pooled analyses and datasets, 55, 112, 123, 126
Population, 88–91, 184
  ELP definition of, 128, 130–134, 137–142, 145–154
  global share, 69–72
  growth, 90, 109, 111, 184–185
  lateral pressure, 42–43, 54–55, 58, 60, 63
Positive feedback, 25, 61, 134
Positivists, legal, 226n53
Possibilism, 20–21
Post–Cold War era, 47, 62–63
Post–Rio de Janeiro era, 67
Postwar (i.e., WWII) era, 13, 14, 158
Post–Cold War era, 47, 62–63
Power-based analysis, 16–22

Prediction, possibility of, 8, 135–136, 170
Price, 11, 26, 81, 84–86, 103, 125, 135, 157, 167, 205
Prigogine, I., 11, 173, 176, 219
Probabilism, 20–21
Production, 2–3, 7, 9, 50–52, 82, 85, 102, 104, 109, 128–131, 161, 167, 215, 223n26
Profile, lateral pressure definition of, 42, 55, 60, 69–72, 177–187
Profits, search for, 48, 102, 104, 141, 217
Property rights, 168
Protectionism, 8, 13, 51, 104
Protest. See Seattle riots
Proximity, model representation of, 20, 55–56, 111
production and consumption, 102, 117
Proxy variables, 60, 69, 79, 137, 203, 205
Psychologists, experimental, 25
Purse-seine nets, 66
Putnam, R. D., 40, 124
Puzzle, research, 30

Qatar, 70, 91, 109, 179, 189, 223–224n28
Quantitative revolution, 18, 22

R-squared (R^2) analysis, 191–195
Randers, J., 222n20, 226n45
Rand McNally, 77
Ratio, trade, 86, 88, 183
Rationality, 37, 40, 42, 48
Rationality, bounded. See Bounded rationality
Ratzel, F., 17–18, 20
Realism, 42, 63, 117. See also Neorealism
Reddy, R., 172
Reductionism, 176
Reference mode, 222n20
Reformers, social, 226n23

Regime, 33–34, 104
Regional analysis, 3, 30–31, 54–56, 78, 90, 107, 110–111, 117, 120, 124–126, 135, 141, 149, 152, 177
Regression analysis, 106–124, 191–200
Regulation, 9, 49, 100–101
Reinforcing relationships. See Positive feedback
Relationships, causal, definition of, 23–26, 133–134
Religion, 7, 20
Renouvin, P., 16–18, 20
Residual-fit analysis, 191–195
Resource trade, 139–140, 143–144, 151–152, 204, 215
Resources, per lateral pressure, 32, 40–43, 45, 47, 50–51, 55, 58–60, 63, 128, 130–131, 138–153, 156–159, 204
Ricardo, D., 10, 12, 127–131, 136, 146–147, 153
Ricardo-Viner trade model, 51, 205, 215
Richards, J. F., 92
Richardson, J. M., 136
Richardson, G. P., 221n8
Rio de Janeiro Earth Summit, 66–67
Riots. See Seattle riots
Rivalry, 41
Roberts, C., 12
Rodrik, D., 8–9
Rowell, J., 27
Rucker, R., 173–174, 223n23
Ruggie, J. G., 38–39, 221n9
Russia, 91, 180, 223–224n28. See also Soviet Union
Rwanda, 70, 91, 109, 189, 223–224n28

S-PLUS™, 191, 225n38
Sagan, C., 32
Samuelson, P. A., 11–12, 128, 130, 148, 207, 213
Sanctions, 32

Satellite photography, 92
Saudi Arabia, 70, 167, 183, 223–224n28
Saurin, J., 67, 97
Savings and loan crisis, 102
Scenarios, ELP
  with trade, 147–148
  without trade, 145–146, 219–220
Schneider, S. H., 96
Schol, E., 172
Scientific method, 221n6
SD. *See* System dynamics
Seattle riots, 1–4, 9–10, 15, 27, 155–158, 166–169
Semi-feudal society, 11
Shiva, V., 7
Short-term economic gain, 8, 16, 48, 165–169
Shrimp. *See* Turtle and shrimp trade dispute
SID (Social Indicators of Development). *See* World Bank
Significance, statistical, 99, 106, 112, 114, 122, 124–125, 191–200
Simon, H. A., 20, 25, 39, 42, 47, 57–58, 222nn10, 13
Simulation, 23–27, 127–154, 219–220
Smith, A., 10–12, 128
SO_2. *See* Sulfur dioxide
Social environment. *See* Environments, natural and social
Social Indicators of Development (SID). *See* World Bank
Solaris™, Sun, 225n225
Soroos, M. S., 33
South, definition of, 66, 70–74, 106–107
Southeast Asia, 30, 58, 70, 110, 121–122, 190
Sovereignty, 20, 37, 102, 115
Soviet Union, 14, 18–19, 44, 61–62, 65, 79, 91, 180. *See also* Russia
SPARC™, Sun, 225n38
Sparta, 61
Spatial analysis, 68–77, 116–124

Spatial autocorrelation. *See* Autocorrelation
Specialization, 11–12, 82, 100, 130
Species loss, 74
Spiral, mathematical definition of, 173
Sprout, H., 20, 222n12
Sri Lanka, 70, 191, 195, 223–224n28
Starr, H., 20–21, 222n12
State. *See* Nation-state
Statistical analysis, 67, 99–126
StatSci, 179
Stella. *See* HPS
Sterman, J. D., 23, 61, 136, 165
Stewart, I., 161, 173
Stock-flow relationships, 23–25, 131–132, 164
Stockholm, 65
Strong, M. F., 49
Subnational data, 78
Sub-Saharan Africa, 7, 30
Sulfur dioxide (SO_2), 103, 106
Summers, R., 80–81
Superpowers (U.S. and Soviet Union), 19, 62, 66
Sustainability, 14–15, 48, 92, 104, 127, 146, 152, 165, 219–220
Switzerland, 70, 179, 221n2, 223–224n28
Sykes, C., 221n6
Synchronization, 30–31, 58, 149, 152, 160, 163
System dynamics, 54–58, 127–154
Systemic dimensionality. *See* Dimensionality, systemic representation of

t-ratio. *See* Significance, statistical
Tariff, 2, 155, 158–159, 169
Taylor, P. J., 18, 21–22
TC × GNP. *See* Trade Connected GNP
Technology, 42–43, 46–47, 55, 59–60, 63, 69, 73, 77, 137, 140
Temporal autocorrelation. *See* Autocorrelation

Thailand, 70, 169, 223n26, 223–224n28
Thermonuclear war, 62
Third wave theory, 34
Third World, 14, 65
Thompson, J., 174–175
Thucydides, 61–62
Time-frame of study, 74, 76, 92, 105, 147, 149, 219, 224n31
Time-series analysis, 78–95
Tokyo, 102
Tolba, M. K., 65
Trade, 85–88
  ELP definition of, 59, 127–131, 136–141, 143–144
  ratio (see Ratio, trade)
  resource, 139, 204
  technical, 140, 206–207
Trade Connected GNP (TC × GNP), 99, 125, 156–157, 162–163, 166
  consequences of, 121–124
  development of, 117–121
  testing of, 191–200
Transportation technology, 50, 82, 85–86, 96
Trinidad and Tobago, 70, 179, 224
Truth, the, 149
Tuna and dolphin trade dispute, 66, 101
Turner, P., 115
Turner, B. L., 81–82, 85, 88, 96, 223n24, 224n30
Turtle and shrimp trade dispute, 223n26
Tutsis, 109
Two-level games, 124
Two staged least squares (2SLS), 43, 56

Ullman, R., 65
UNCED (United Nations Council on Environment and Development). See Rio de Janiero
Uncertainty, 20, 50, 74
Underdevelopment, 152

UNFAO (United Nations Food and Agriculture Organization), 75, 92, 189
Unilateral international action, 2, 66
Unintended consequences, 7, 37, 39, 52, 58, 82, 85–86, 94–96, 169–170
United Kingdom. See Great Britain
United Nations, 14, 65–66, 100
United States 2, 6, 8, 12–14, 44, 61, 65–66, 70, 79, 86, 101, 108, 120, 140, 146, 166–167, 169, 171, 177, 179–184, 198, 213, 216, 223–224n28
Univariate analysis, 106–111
UNIX™, 225n38

Validation, 61
Variables of ELP model, 202–209
Variance, explanation of, 191–195
Venezuela, 70, 167, 179, 180, 223–224n28
Vertical differentiation, 21–22, 59
Vietnam, 19
Visualization, 37, 55, 69, 201

Wallerstein, I., 21
Walt, S. M., 65
Waltz, K. N., 35–41, 45, 52, 61, 81, 83, 159
War, study of, 13, 19–20, 31, 35–44, 52, 86, 158–159
Ward, M. D., 18–19, 22, 225n35, 226n55
WEF. See World Economic Forum
Weiss, E. B., 33
Welfare, human, 85, 101, 159, 224n32
Western Europe, 121–122, 167, 190
Whitmore, T. M., 30, 46, 89, 93
Williams, M., 66, 68, 92
Willingness. See Opportunity and willingness framework
Wilmsen, S. K., 102
Wils, A., 168
Wilson, E. O., 30

Wilson, W., 13
Woo, W. T., 27
World Bank, 3, 13, 71–72, 79, 89–96, 158, 189, 223–224n28, 224n31, 225n43
World Economic Forum (WEF), 3, 221
World Resources Institute, 75
World Trade Organization (WTO), 1–4, 7, 9, 15, 27, 100, 155, 158, 165, 223n26
World War I, 13, 19, 43, 146
World War II, 13–14, 18–19, 43–44, 49–50, 53–54, 80, 82, 89, 91
WTO. *See* World Trade Organization

Xerox, 6

Young, O. R., 33
Yugoslavia, 120

Zaire, 109